HARNESSING SCIENCE
FOR
ENVIRONMENTAL
REGULATION

Harnessing Science
for
Environmental
Regulation

EDITED BY
John D. Graham

PRAEGER

New York
Westport, Connecticut
London

Library of Congress Cataloging-in-Publication Data

Harnessing science for environmental regulation / edited by John D.
  Graham.
      p. cm.
   Includes bibliographical references and index.
   ISBN 0–275–93766–6 (alk. paper)
   1. Environmental law—United States. 2. Hazardous substances—Law
and legislation—United States. 3. United States. Environmental
Protection Agency. Science Advisory Board. 4. Chemical Industry Institute
of Toxicology. 5. Health Effects Institute. 6. Pollution—Health
aspects—Research—United States. 7. Hazardous substances—Health
aspects—Research—United States. I. Graham, John D. (John David),
1956– .
KF3775.H365   1991
344.73′046—dc20
[347.30446]   90–47331

British Library Cataloguing in Publication Data is available.

Library of Congress Catalog Card Number: 90–47331
ISBN: 0–275–93766–6

First published in 1991

Praeger Publishers, One Madison Avenue, New York, NY 10010
An imprint of Greenwood Publishing Group, Inc.

Printed in the United States of America

The paper used in this book complies with the
Permanent Paper Standard issued by the National
Information Standards Organization (Z39.48–1984).

10 9 8 7 6 5 4 3 2 1

CONTENTS

PREFACE

This book is an outgrowth of the deliberations of the Belmont Workshop on Scientific Advice for Environmental Health Regulation. This small invitational workshop, which was sponsored by Harvard University, was convened in October 1988 at the Belmont House in Elkridge, Maryland. A list of workshop participants is given in the Appendix.

The primary source of funding for the book was a cooperative agreement between the U.S. Environmental Protection Agency (EPA) and the Harvard School of Public Health. Supplemental research funding was also made available by the Andrew Mellon Foundation. The Belmont workshop itself was funded by the American Industrial Health Council. The case study on carbon monoxide (Chapter 9), which was added to the book after the Belmont workshop, was supported under a separate grant to Harvard University by the Motor Vehicle Manufacturers Association. Near the end of the project, Dow Chemical Company and Monsanto Company made small grants to Harvard that allowed the book to be completed. I thank all of these organizations for their financial assistance.

The book evaluates three ''regulatory science'' organizations: The EPA Science Advisory Board, the Chemical Industry Institute of Toxicology, and the Health Effects Institute. Readers should recognize that the book covers the activities of these organizations only through 1988, before each of them underwent significant changes in leadership.

Any effort of this magnitude has numerous contributors whose names may not appear as authors of particular chapters. I received a tremendous amount of help from John Evans, Fred Hoerger, Donald Hornig, Sheila Jasanoff, William Lowrance, Roger McClellan, Paul Portney, James Senger, and James Wilson. These individuals went beyond offering constructive criticism and provided me with personal encouragement at various times when I needed it. For preparing the manuscript itself, I am indebted to Marilyn Graham and Patricia Worden. Obviously, I am very grateful to each of the authors of the chapters for providing their contributions on such a timely basis.

HARNESSING SCIENCE
FOR
ENVIRONMENTAL
REGULATION

CHAPTER 1

SCIENCE AND ENVIRONMENTAL REGULATION

John D. Graham

Many American citizens are frightened about the prospect of being exposed to toxic substances that may cause various forms of cancer, birth defects, and other adverse health effects. While no one knows the overall magnitude of the public health damage caused by toxic chemicals, episodes of chemical pollution at Times Beach, Missouri, Love Canal, New York, and Bhopal, India, have led to highly publicized human tragedies. Through the mass media the public is told that potentially hazardous chemical exposures occur in daily life—through the food we eat, the water we drink, and the air we breathe. Consequently, an elaborate regulatory system has been established in the United States to protect people from such exposures.

Science plays an essential role in toxic chemical regulation. The regulatory system cannot work effectively without sound scientific data and thoughtful scientific judgment. In particular, regulators need science to help them distinguish the serious from the trivial exposures to toxic chemicals. They also need science in their efforts to devise and defend protective regulations.

While science plays an essential role in protective regulation, the norms of science can obstruct regulatory progress and threaten some of our cherished democratic values. How could the institution of science, that honorable search for truth, possibly have these pernicious effects? It is now well recognized that the inherent conservatism of science, that a good scientist does not allow commitment to a hypothesis until the case is proven, can work against the prudence embedded in most laws and regulations drafted to protect public health. As long as the burden of proving risk to human health lies with the regulator, any imperfections in scientific knowledge about human risk can operate to permit continued human exposures to toxic chemicals.[1] Moreover, a sincere attempt by regulators to seek advice from scientists can result, albeit quite subtly, in a transfer of regulatory power from accountable politicians to unaccountable experts.[2]

Regulations of chemical production, use, and disposal are usually made in the presence of scientific uncertainty and political conflict. The economic stakes in such decisions can be enormous. Value judgments must be made about how much health protection is feasible and affordable and who should pay the costs of cleanup. In our representative democracy the American citizen has a right to hold politicians accountable for the consequences of these decisions. To argue otherwise is to contest the virtue of self-governance and invite a tyranny of experts. In the words of Thomas Jefferson: "I know no safe depository of the ultimate powers of the society but the people themselves; and if we think them not enlightened enough to exercise their control with a wholesome discretion, the remedy is not to take it from them, but to inform their discretion."[3]

The challenge, then, is to design a regulatory process that captures the knowledge of science while safeguarding the proper domain of accountable political choice. In this book we examine a variety of institutional strategies for meeting this challenge. Focusing on the U.S. Environmental Protection Agency (EPA), we examine the regulatory roles played by the EPA Science Advisory Board, the Chemical Industry Institute of Toxicology, and the Health Effects Institute. While each of these organizations is unique in its origin, mission, and evolution, the book argues that each institution is making a significant contribution to resolving America's "regulatory science" dilemma.[4] The goals of the book are to describe how and why these organizations were created, to examine what roles they play in practice, and to recommend how they might work better in the future.

THE RISE OF REGULATORY POWER

The 1960s and 1970s were an era of increasing public support for regulation of private enterprises to protect human health, safety, and the environment. The support for regulation was galvanized by Ralph Nader and the emerging environmental movement. Throughout the 1970s the U.S. Congress passed new laws that provided the federal government with expansive regulatory authority over the production of chemicals, their use, and their discharge into the environment.[5] These laws included the 1970 and 1977 amendments to the Clean Air Act, the Water Pollution Control Act of 1972, the Federal Insecticide, Fungicide, and Rodenticide Act of 1972, the Safe Drinking Water Amendments of 1974, the Toxic Substances Control Act of 1976, the Resource Conservation and Recovery Act of 1976, and the 1980 Superfund law governing cleanup of hazardous waste sites. Within the executive branch the EPA was created in 1970 and became the primary agency responsible for implementing these new mandates.

The regulatory power created in the 1970s was remarkably different from the old-style economic regulation that emerged from the New Deal era.[6] The old-style regulation was based on the premise that imperfections in the marketplace could be solved by granting expert administrative agencies power to control the amount of industrial competition. Economic regulators were authorized by the Congress to pursue "the public interest" without specific deadlines or prescriptions. For example, Congress trusted the Interstate Commerce Commission to exercise its powers over rates and entry into the railroad industry in an expert fashion. In contrast to this approach, the environmental statutes of the 1970s were more expansive. They provided the EPA with diverse powers over multiple industries. At the same time, the degree of faith in expert administrative discretion has diminished. Environmental statutes—especially the more recent ones—often have an "agency-forcing" character.[7] For example, environmental statutes may compel EPA to take specific regulatory steps within fixed deadlines. Even when EPA is granted significant discretion, the law often specifies what factors or types of evidence may be considered and what procedures must be followed. The new-style regulation also invites citizen suits and judicial review to assure that the legislative mandate has been faithfully executed.[8]

Legislators were concerned that scientific uncertainty about the effects of exposure to toxic chemicals might slow the pace of regulation at EPA. To address this problem, Congress made clear that it wanted EPA to act in a preventive fashion—to order reductions in human exposures to toxic chemicals even if the harmful effects of such exposures were not proven definitively. The authority to act without conclusive science proved to be quite controversial.

In one of the precedent-setting cases under the Clean Air Act, *Ethyl Corp. v. EPA*, a federal appeals court upheld EPA's precautionary approach to setting air quality goals for lead pollution. Judge Skelly Wright wrote that the EPA may draw conclusions from "suspected, but not completely substantiated, relationships between facts, from theoretical projections from imperfect data, from probative preliminary data not yet certifiable as 'fact,' and the like."[9] The scientific basis of the lead rules was in fact much stronger than EPA tends to have in most regulatory proceedings.

In several related cases dealing with EPA regulation of suspected human carcinogens, the federal courts ruled that Congress foresaw that EPA might need to regulate on the basis of carcinogenicity data in laboratory animals, since it might be too dangerous to await scientific proof of cancer risk in humans.[10] Where the public health is potentially at stake, Congress and the federal courts decided in the 1970s that EPA is expected to regulate in the face of scientific uncertainties.

Although Congress sought a prudent regulatory approach, the federal courts have also pointed out that Congress expects that regulators will make decisions that can pass some minimum test for rationality. Under the Administrative Procedures Act, Congress authorized federal courts to block any agency decision that was "arbitrary, capricious, an abuse of discretion or otherwise not in accordance with the law."[11] Under some circumstances federal courts also require ''substantial evidence'' in support of the factual findings that support an agency's decision.[12] These doctrines are not easy to apply when the scientific basis of regulation is complex and uncertain.

In the context of disputes about toxic chemical regulation, the courts are still wrestling with the question of how much scientific analysis is necessary to support a rulemaking. The only Supreme Court decision on this question, which arose out of a dispute about occupational exposures to benzene, suggests that federal courts should expect quantitative risk assessments when reviewing agency decisions.[13] More recently, a federal appeals court held that a federal agency's risk assessment of formaldehyde was insufficient to justify a ban on certain applications of urea-formaldehyde foam insulation.[14] The agency had made some pessimistic assumptions about the extent of human exposure to formaldehyde and about the relative sensitivity of rats and humans. While both of these federal court decisions were quite controversial, it is apparent that judges will not tolerate regulatory risk assessments that do not have some degree of scientific support.

SCIENTIFIC AND REGULATORY INCOMPETENCE

By the end of the 1970s, America's track record with environmental health regulation was not entirely satisfying. Critics of EPA in particular argued that environmental regulation was hurting the national economy by aggravating inflation, impairing industrial productivity, discouraging technological innovation, and disadvantaging the competitive position of domestic firms in the global economy.[15] Several economists estimated that environmental regulations were costing Americans $80 to $100 billion per year—costs that were especially painful during a period of high energy prices, double-digit inflation, and escalating interest rates.[16]

The large costs of environmental regulation might have been considered acceptable if the public had acquired a sense of confidence in EPA's ability to use science to tackle important public health problems. Many EPA initiatives have in fact been shown to have a strong scientific basis, and several EPA regulations have resulted in substantial health, ecological, and economic benefits.[17] Economists, for example, have estimated that the benefits of marketing unleaded gasoline to consumers are much greater than the costs of producing unleaded gasoline. These successes, however, have been offset in the public eye by a series of damaging instances of scientific and regulatory incompetence. In short, EPA and other federal regulatory agencies have acquired a reputation for not always developing and using good science in their regulatory decisions.

Many examples can be cited where significant regulatory decisions were made on the basis of flawed, distorted, misinterpreted, or nonexistent science. Some examples follow.

• The Food and Drug Administration's proposals in the late 1970s to ban saccharin and nitrates were ultimately discredited, the latter because of revelations that a key laboratory experiment had been misinterpreted and the former due to both public protest and credible suggestions that the health risks were not as large as FDA scientists had predicted.[18]

- EPA's original plan to ban the herbicide 2,4,5-T came under fire when the key health study—which purported to show an association between spraying activities and miscarriages in the state of Oregon—could not survive independent scientific review.[19]

- EPA's rationale for its primary ambient air quality standard for carbon monoxide was undercut twice in the 1970s: once when the key studies on psychomotor effects could not be replicated and again when two key studies of angina patients were found to have inadequate documentation of critical data.[20]

- EPA's large-scale research program on the effects of air pollution on child health was damaged and ultimately dismantled when agency officials were accused of overinterpreting the results in order to provide ammunition for regulatory initiatives.[21]

- EPA's handling of the Love Canal episode generated widespread criticism when an EPA-supported study of chromosomal abnormalities was leaked to the press prior to scientific peer review. The study, which had been hastily commissioned to support the agency's cleanup strategy, was ultimately judged to be inconclusive due to the lack of an appropriate control group.[22]

- In the late 1970s a team of government scientists circulated a draft report that contained an estimate that 20 to 40 percent of all cancers in the United States might be caused by occupational factors. The report, which was never published, received widespread media attention and was cited as evidence to support new regulatory initiatives. In the face of severe scientific criticism, several well-known scientists denied that they had authored the report, even though they were cited as authors.[23]

Each of these examples could be expanded into an extensive discussion. Regardless of the merits of each situation, there is no doubt that the cumulative effect of these incidents was to weaken the scientific credibility of federal regulation.

The credibility problem was exacerbated by political appointees in the Reagan administration who pursued the president's philosophy of "regulatory relief" without careful scientific analysis. While numerous examples can again be cited, several should suffice.

- A decision not to give regulatory priority to formaldehyde under the Toxic Substances Control Act was justified in a poorly reasoned memorandum that was never subjected to scientific peer review. The decision was later reversed after widespread criticism from environmentalists, scientists, and members of Congress.[24]

- A decision by the Occupational Safety and Health Administration (OSHA) not to regulate ethylene oxide—a known human carcinogen—was contested in federal court by labor unions and public health organizations. The court ordered OSHA to prepare a reasoned risk assessment and begin the rulemaking process.[25]

- The EPA spent years debating how to quantify the cancer risks associated with ethylene dibromide (EDB) despite sound scientific evidence of carcinogenicity in animals and significant consumer exposures to EDB residues in fruit and grain products. Although some uses of EDB on grain were suspended in early 1984, EPA was widely criticized for moving too slowly in response to scientific evidence.[26]

• The Reagan administration came under harsh criticism from numerous quarters for permitting the Office of Management and Budget (OMB) to delay and block new regulatory initiatives. Critics pointed out that the OMB's regulatory review staff was comprised primarily of economists. There were no toxicologists, epidemiologists, or health scientists at OMB to review EPA proposals.[27]

• In an attempt to purge EPA of the influence of scientists with proregulation inclinations, the first EPA administrator in the Reagan administration, Anne Gorsuch, permitted distribution of a list of scientists who should not be retained on EPA advisory committees. Many of these scientists had their advisory appointments terminated. When William Ruckelshaus returned to EPA in 1983, many of these advisory appointments were renewed.[28]

These kinds of incidents further eroded public confidence in the ability of EPA and other executive-branch agencies to perform objective scientific analyses of toxic chemical issues.

CREATIVE STRATEGIES FOR HARNESSING SCIENTIFIC EXPERTISE

The problem of incompetent regulatory science has not been ignored. During the last fifteen years numerous efforts have been made to strengthen the scientific basis of toxic chemical regulation. These efforts include new approaches to generating scientific data as well as new approaches to interpreting scientific knowledge for regulatory purposes. Most of these efforts, although serious, were ad hoc attempts to strengthen the hand of science within existing institutional arrangements. Examples of such efforts include the cancer risk-assessment guidelines published by the Interagency Regulatory Liaison Group (IRLG) in the Carter administration and the "cancer principles" published by the White House Office of Science and Technology Policy in the Reagan administration.[29] Numerous committees of the National Academy of Sciences have been formed to provide scientific judgments on important regulatory questions. Regulatory agencies and many corporations have also made efforts internally to produce more science and assist in its interpretation for regulatory purposes. The American Industrial Health Council, an industry-supported trade organization, was created in 1978 to advocate the use of science in regulatory decision making and has been a persistent voice in favor of better regulatory science.

As important as these efforts have been, this book focuses on several efforts that embody more radical departures from traditional institutional arrangements. The Scientific Advisory Board (SAB) of EPA was authorized by the Congress in 1978 to provide the EPA administrator with independent, expert advice on a wide range of issues. SAB operates through the service of scientists on various topical committees and subcommittees. The Chemical Industry Institute of Toxicology (CIIT) was created in 1976 by a consortium of petrochemical companies that recognized the need for an independent scientific organization to test chemicals for toxicity and develop new methods for assessing toxicity. While CIIT had no formal affiliation with any government agency, its sponsors understood that CIIT science might become the basis for regulatory decisions. The Health Effects Institute (HEI) was created in 1980 to produce and interpret scientific data on health effects that are relevant to making regulatory decisions about pollution from mobile sources. HEI is jointly and equally funded by the EPA and a consortium of engine and vehicle manufacturers. Although these three organizations are not the only new entities that have been created to address regulatory science issues, each is important in its own right, and they provide enough diversity to make it possible to consider the implications of different models of institutional innovation.[30]

This book explores a series of fundamental questions about the nature of these organizations. How and why were they created? How are they perceived by various actors in the regulatory process? How do they cope with the tension between pursuing scientific values and pursuing society's interests in controlling human exposure to chemicals? What kinds of influence have these organizations exerted in specific regulatory controversies? How well have these organizations accomplished their respective missions? Insofar as these organizations make a constructive contribution, how can they and others be nurtured, strengthened, and more widely utilized? These are important questions to consider because such organizations, while small in size, are becoming a central part of what Sheila Jasanoff has termed "the fifth branch of government."[31]

ORGANIZATION OF THIS BOOK

Chapters 2 through 4 were written by former leaders of SAB, CIIT, and HEI, respectively. These chapters are intended to provide background information on these organizations as well as some historical perspective by individuals who have led these new institutional entities. Chapters 5 through 9 are case studies in regulatory science. The specific substances covered in the case studies are unleaded gasoline, perchloroethylene, formaldehyde, nitrates, and carbon monoxide. The case studies are historical accounts of how one or more of these organizations participated in a regulatory controversy. Although the case studies are not necessarily representative of each organization's entire history, they do cover some of the most significant activities of each organization. The final chapter examines how well these organizations have performed in the past and how they might better serve society in the future. Information for the final chapter comes not merely from the case studies and the author's firsthand experiences with the organizations, but also from discussions at an October 1988 workshop sponsored by the Harvard School of Public Health. The Appendix at the end of the book lists the participants in the workshop.

NOTES

1. H. Latin, "The 'Significance' of Toxic Health Risks: An Essay on Legal Decisionmaking under Uncertainty," *Ecology Law Quarterly*, vol. 10, 1982, p. 339; H. Latin, "Good Science, Bad Regulation, and Toxic Risk Assessment," *Yale Journal on Regulation*, vol. 5, 1988, pp. 89–148.

2. J. D. Graham, L. Green, and M. J. Roberts, *In Search of Safety: Chemicals and Cancer Risk*, Harvard University Press, Cambridge, Mass., 1988.

3. Thomas Jefferson, quoted in D. L. Bazelon, "Risk and Responsibility," *Science*, vol. 205, 1979, p. 278.

4. The phrase "regulatory science" is used by Sheila Jasanoff to encompass knowledge production, knowledge synthesis, and prediction for use by actors in the regulatory process. Sheila Jasanoff, *The Fifth Branch of Government*, Harvard University Press, Cambridge, Mass., 1990, chapter 3.

5. Committee on Risk and Decision Making, *Risk and Decision Making*, National Research Council, Washington, D.C., 1982.

6. W. Lilley and J.C. Miller, "The New 'Social' Regulation," *Public Interest*, vol. 47, 1977, pp. 49–61.

7. J. D. Graham, "The Failure of Agency-Forcing: The Regulation of Airborne Carcinogens under Section 112 of the Clean Air Act," *Duke Law Journal*, 1985, pp. 100–150.

8. R. Shep Melnick, *Regulation and the Courts*, Brookings Institution, Washington, D.C., 1983.

9. Ethyl Corp. v. EPA, 541 F.2d 1 (D.C. Cir. 1976).

10. EDF v. EPA, 465 F.2d 528 (D.C. Cir. 1972).

11. 5 U.S.C. 551.

12. Industrial Union Department, AFL-CIO v. Hodgson, 499 F.2d 467, 469–473 (D.C. Cir. 1974).

13. Industrial Union Department, AFL-CIO v. Hodgson, 448 U.S. 607 (1980).

14. Gulf S. Insulation v. CPSC, 701 F.2d 1137 (5th Cir. 1983).

15. P. W. MacAvoy, *The Regulated Industries and the Economy*, W. W. Norton and Co., New York, 1979.

16. R. W. Crandall, *Controlling Industrial Pollution*, Brookings Institution, Washington, D.C., 1983.

17. K. L. Florini, G. O. Krumbhaar, Jr., and Ellen K. Silbergeld, "Legacy of Lead: America's Continuing Epidemic of Childhood Lead Poisoning," Environmental Defense Fund, Washington, D.C., March 19, 1990.

18. R. J. Smith, "Nitrates: FDA Beats a Surprising Retreat," *Science*, vol. 209, 1980, pp. 1100-1101; F. B. Cross, *Environmentally Induced Cancer and the Law*, Quorum Books, New York, 1989, p. 118.

19. M. Gough, *Dioxin, Agent Orange*, Plenum Press, New York, 1986.

20. U.S. EPA, *Review of the NAAQS for Carbon Monoxide: Reassessment of Scientific and Technical Information*, EPA–450/5–84–004, July 1984.

21. U.S. House of Representatives, *The Environmental Protection Agency's Research Program with Primary Emphasis on the Community Health and Environmental Surveillance System (CHESS): An Investigative Report*, Washington, D.C., November 19, 1976.

22. G. Kolata, "Love Canal: False Alarm Caused by Botched Study," *Science*, vol. 208, 1980, pp. 1239–42.

23. R. Peto, "Distorting the Epidemiology of Cancer: The Need for a Balanced Overview," *Nature*, vol. 284, 1980, pp. 297–300.

24. N. Ashford, C. W. Ryan, and C. C. Caldert, "Law and Science Policy in Federal Regulation of Formaldehyde," *Science*, vol. 222, 1983, p. 893.

25. Public Citizen Health Research Group v. Auchter, 554 F. Supp. 242 (D.D.C.) affirmed, 702 F.2d 1150 (D.C. Cir. 1983).

26. Cross, *Environmentally Induced Cancer*, p. 114.

27. Eric D. Olsen, "The Quiet Shift of Power: Office of Management and Budget Supervision of Environmental Protection Agency Rulemaking under Executive Order 12,291," *Virginia Journal of Natural Resources Law*, vol. 4, 1984, p. 1.

28. U.S. EPA, "Science Advisory Board 'Hitlist,' " unpublished memorandum, 1982.

29. Interagency Regulatory Liaison Working Group on Risk Assessment, "Scientific Bases for Identification of Potential Carcinogens and Estimation of Risks," *Federal Register*, vol. 44, 1979, p. 39858; Office of Science and Technology Policy, "Principles for Cancer Risk Assessment," *Federal Register*, vol. 49, 1985, p. 21593.

30. Other scientific advisory groups to regulatory agencies include the Clean Air Scientific Advisory Committee (CASAC) of EPA, the Science Advisory Panel (SAP) of EPA, the Chronic Hazard Advisory Panels at the Consumer Product Safety Commission, the Technical Advisory Committees at the Food and Drug Administration (FDA), and the Agency for Toxic Substances and Disease Registry at the Centers for Disease Control. See T. S. Burack, "Of Reliable Science: Scientific Peer Review, Federal Regulatory Agencies, and the Courts," *Virginia Journal of Natural Resources Law*, vol. 7, 1987, pp. 27–110.

31. Jasanoff, *Fifth Branch of Government*.

CHAPTER 2

THE EPA SCIENCE ADVISORY BOARD

Terry F. Yosie

Institutions, like people, pass through developmental cycles. Institutions have the capacity to renew themselves if they adapt to changing external circumstances by redefining their mission and managing their human and other resources to take account of change. The premise of this chapter is that individual units within institutions also experience these cycles, and that their role and influence can wax and wane depending upon external events and internal leadership. Using the Science Advisory Board (SAB) of the U.S. Environmental Protection Agency (EPA) as a case study, this chapter discusses the evolution of the board's role, assesses the formalization of its review activities, identifies prerequisites for maintaining an effective and continuing independent scientific advisory process within regulatory agencies, and, finally, identifies some potential new directions for the SAB and for science within the context of a changing EPA.

MISSION, STRUCTURE, AND EVOLVING ROLE OF THE SAB

The mission of the SAB has remained relatively constant over the past fifteen years, although the scope of its activity and the means by which this mission is implemented have significantly changed. The Congress codified the board's mission in 1978 through the passage of amendments to the Environmental Research and Development, Demonstration Authorization Act (ERDDAA). The act called for the SAB to perform three kinds of tasks:

• To "provide such scientific advice as may be requested by the Administrator" or by Senate and House committees having oversight over EPA activities.

- To conduct an annual review of the scientific adequacy of EPA's five-year research and development plan. In practice, this requirement has evolved into a broader oversight of the scientific adequacy, needs, and direction of EPA research in general.

- To have the opportunity to review the technical adequacy of "any proposed criteria document, standard, limitation, or regulation" developed under EPA's authorizing statutes.[1]

Upon its formation in the early 1970s the board was located in EPA's Office of Research and Development. In 1976 it became a staff office within the Office of the Administrator to provide it with more independence and expand its scope to the entire agency. ERDDAA required that SAB report directly to the administrator.

ERDDAA, upon its passage, stated that the SAB "is authorized to constitute such member committees and investigative panels as the Administrator and the Board find necessary to carry out" its statutory responsibilities.[2] Over time, the board developed a stable, yet flexible, structure to address the growing range of issues submitted for its review. This structure consisted of three major components: an executive committee that functioned as a board of directors, meeting quarterly to review SAB committee reports, authorize new scientific reviews, and provide oversight for SAB operations; standing committees that conducted the majority of SAB reviews, prepared reports, and recommended options for executive committee consideration; and ad hoc subcommittees that performed the same function as standing committees but reviewed more specialized issues not within the scope of the standing committees' jurisdiction.[3] Ad hoc subcommittees were generally abolished upon completion of their assignments. Figure 2.1 illustrates the SAB structure in fiscal year 1990.

There is a growing body of literature that describes and assesses the board's role and performance. This literature can be subdivided into at least three categories: case studies of issues reviewed by the SAB; evaluations of the development and use of scientific information/risk assessment within EPA and the board's contribution in this context; and the evolutionary role of peer review in regulatory agencies. The reference section at the end of the chapter cites a selected sample of this literature.

Several sets of factors help explain the board's expanded activity. These include the following:

- Transformation of EPA's task at its inception (responding to a perceived sense of crisis for well-defined problems in the relative absence of science) to a differing role in later stages of its history (developing and implementing regulations that require scientific and technical support)

- Periodic challenges to EPA's scientific credibility

- Risk assessment and scientific review as legitimizers of policy choices

- Desire of SAB members to influence EPA decision making.

Figure 2.1
Science Advisory Board: Committee Structure

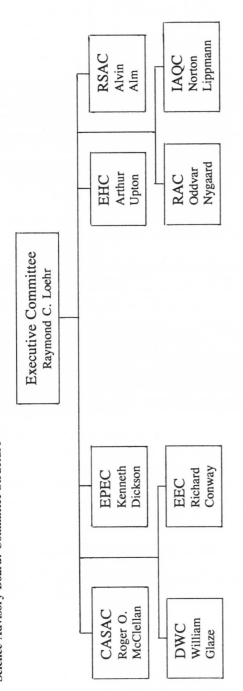

CASAC = Clean Air Scientific Advisory Committee
DWC = Drinking Water Committee
EEC = Environmental Engineering Committee
EHC = Environmental Health Committee
EPEC = Ecological Processes and Effects Committee
IAQC = Indoor Air Quality/Total Human Exposure Committee
RAC = Radiation Advisory Committee

In general, the SAB's history unfolded in at least three phases during the years 1973-1988. These phases, the principal function(s) performed by the board, and major EPA and external events shaping the SAB's experience are presented in Table 2.1.

FORMALIZATION OF THE SAB REVIEW PROCESS

Over time, the SAB review process became more formalized as EPA science and policy administrators responded to particular scientific controversies and sought the aid of external scientists to rebuild or improve the agency's scientific credibility. This formalization was greatly advanced through particular SAB reviews or through instances where EPA's scientific credibility was challenged because of the lack of sufficient scientific consensus concerning the agency's interpretation and use of particular scientific studies. Key events that propelled this formulation are discussed here.

CHESS: Recognition of the Need for Independent Scientific Review

In the early 1970s, following the formation of the EPA, the agency set a number of air pollution standards to reduce human exposures. At the same time, it undertook a major series of epidemiological investigations in a number of cities to better document air pollution concentrations and their relationship to human health. The results of one set of such studies, known as the Community Health and Environmental Surveillance System (CHESS), were intended to support future standard setting for such pollutants as nitrogen oxides, particulate matter, and sulfur oxides. The program operated from 1970 to 1975, and a monograph of the results was published in May 1974 entitled *Health Consequences of Sulfur Oxide: A Report from CHESS, 1970-1971*. Other papers and presentations on this and other CHESS data sets were subsequently prepared.

The studies' design and execution were subject to a number of scientific flaws.[4] A major shortcoming was the lack of independent peer review to ensure that scientists developing the studies utilized acceptable scientific criteria and would use appropriate caution in applying study results to regulatory decision making. By the mid–1970s the scientific controversy over the CHESS studies within EPA and the scientific community prompted a congressional investigation. George E. Brown, Jr. (D-California), the principal overseer of EPA research in the Congress, solicited the testimony of EPA officials and representatives of the scientific community, including members of the Science Advisory Board. The report of Congressman Brown's investigation was a scathing indictment of EPA's scientific credibility. A major recommendation in the report was to expand the responsibility of the Science Advisory Board in reviewing EPA's research.[5]

There were three principal results of the CHESS experience upon EPA:

Table 2.1

Evolution of SAB History: Phases, Functions, and External Factors

	Major SAB Activities	EPA/External Events
1973 - 1980	- Selected review of research programs - Limited review of assessment documents	- CHESS studies controversy of Congressional investigation - Love Canal study - Controversy over individual assessment documents - IRLG - Limited use of science in regulatory decision making - 1979 and ozone standard API suit
1980 - 1983	- Selected review of research programs - Limited review of assessment documents - Limited review of regulatory issues	- Major reductions of EPA research support - Concerns over politicization of EPA science congressional investigations - "Hit list" controversy - Desire for "good science"
1983 - 1988	- Expanded review of research programs - Expanded review of assessment documents - Codified role in review of regulatory issues - Review of policy statements/ guidance/advisories - Methodology development reviews - Review of specific studies/ surveys - Inter-agency reviews	- Rebuilding of EPA programs - Adoption of risk assessment/risk management framework - Updating and development of risk assessment guidelines - More focused debate on conservatism in risk assessment - Recognition of scientist as an EPA constituency - Science as legitimizer of policy choices

- The CHESS studies were never used as a scientific basis for air pollution standards.
- EPA's epidemiological research program experienced a setback from which it has never recovered.
- There was a direct relationship between the Brown investigation and the passage of the 1978 ERDDAA that created and further empowered the SAB. The Brown investigation was also a motivating factor in Congress's decision to amend Section 109 of the Clean Air Act in 1977 to create an independent scientific committee to review the scientific basis of air quality criteria.

The 1979 National Ambient Air Quality Standard for Ozone: The Closure Principle and Staff Paper Development

Throughout the middle and late 1970s EPA's Office of Research and Development (ORD) and Office of Air Quality Planning and Standards (OAQPS) assessed the scientific evidence that would be used to support revision of the national ambient air quality standard (NAAQS) for ozone.[6] Three events occurred in this process of reassessment that had major implications for EPA's future use of the SAB:

- EPA's use of an air quality criteria document that had not received the consensus support of the SAB review subcommittee as scientifically adequate
- OAQPS's convening of a separate scientific panel to review scientific evidence pertaining to the existing standard and the development of alternative standards
- Initiation of a court suit by the American Petroleum Institute (API) charging that EPA had failed to comply with ERDDAA provisions requiring appropriate "opportunity" for SAB review of the criteria document, thereby resulting in a scientifically unsupportable standard[7]

The API appealed the court case to the Supreme Court. While its case was ultimately denied, API's critique of the process for setting NAAQSs, combined with the criticism of leading members of the scientific community for the manner in which EPA utilized scientific data to set the ozone standard, led to a major reexamination and reformulation of the NAAQS-setting process by EPA professional staff in the years 1979-1981. The SAB's newly formed Clean Air Scientific Advisory Committee (CASAC), while not involved in the ozone-standard review (but having inherited the SAB's statutory responsibility to review air quality criteria), itself undertook an examination of the standard-setting process. The result of these parallel reviews, combined with formal solicitation of public input by CASAC, led to an internal EPA consensus that resulted in two major changes:[8]

- Adoption of the "closure" principle, whereby CASAC would prepare a memorandum to the administrator when it concluded that EPA had prepared a balanced and adequate interpretation of the scientific information for a particular criteria air pollutant.
- Initiation by OAQPS of a "staff paper" whose purpose was to create a bridge document between the multitude of scientific studies contained in the air quality criteria document

and a formally proposed standard. The staff paper identified the key scientific studies that were most relevant to standard setting and proposed "ranges of interest" within which the staff recommended that a standard be established. CASAC review of the staff paper became a routine part of the NAAQS process, including the preparation of a closure memorandum.[8]

On February 8, 1982, EPA Administrator Anne M. Gorsuch, as part of her effort to improve the scientific basis of EPA regulations and standards, formally endorsed the closure and staff paper initiatives.[9]

The 1983 Rulemaking for Airborne Radionuclides: The Principle of Early Scientific Review

Section 117 of the Clean Air Act requires that before listing a pollutant as "hazardous" under Section 112 of the act, the administrator must consult with the agency's scientific advisory bodies.[10] This provision, combined with the ERDDAA language mandating that the EPA provide the SAB with the "opportunity" to review the technical adequacy of proposed regulations prior to their publication, imposed two clear requirements upon the administrator. EPA, in attempting to comply with a Clean Air Act provision that it determine by 1979 whether airborne radionuclides should be listed and regulated under Section 112 of the Clean Air Act, failed to consult with the SAB prior to listing and publishing proposed regulations.

In December 1983 EPA Administrator William Ruckelshaus formally requested SAB review of the scientific basis of the proposed rule.[11] In its subsequent review the SAB subcommittee that evaluated the proposed rule reached several implicit and explicit conclusions. These included the following:

- EPA's Office of Radiation Programs (ORP) existed as a quasi-independent subculture within the agency. This stemmed, in part, from the facts that ORP possessed a separate institutional history than the rest of EPA (its establishment resulted from the breakup of the Atomic Energy Commission); that senior EPA officials generally did not perceive radiation issues as a high priority relative to other pressing public health and environmental problems they had to confront; and that a separate tradition of developing and peer reviewing radiation risk assessments evolved in ORP relative to the rest of the agency.

- Individual elements of ORP's analysis of airborne radionuclide risks were scientifically adequate, but there was no integrating document that enabled either scientists or administrators to explicitly review how scientific information was used as the basis for proposed regulations. As a result, the regulations lacked a scientifically adequate assessment of the health risks from exposure to airborne radionuclides.

- ORP should better integrate its risk-assessment activities with those of other EPA offices. Specifically, ORP should adopt the concept of the "staff paper" that former Administrator Anne Gorsuch had announced as agency policy more than two years earlier.[12]

The most important outcome of this rulemaking as regarding EPA's relationship with the SAB was that it established in the minds of senior officials the importance of early independent scientific review in the rulemaking process. As the Radionuclides Subcommittee stated in its final report to the administrator: "At the time the Subcommittee was asked to begin its review of the scientific data base, the formal rule had already been proposed in the Federal Register. . . . This is not an effective way to involve the Science Advisory Board nor the scientific community in EPA's rulemaking process."[13] On June 25, 1985, EPA Administrator Lee M. Thomas endorsed the statements contained in Anne Gorsuch's 1982 memorandum and explicitly stated that it was EPA policy to seek SAB review early in the rulemaking process.[14]

Development of Risk-Assessment Guidelines, 1984–1986: Scientific Review as Legitimizer of Science-Policy Choices

By the mid–1980s there was wide recognition within EPA and the scientific community of the need to formally and explicitly state how EPA would practice risk assessment in regulatory decision making. This recognition was driven by several factors: the growing use of risk assessment as a major, if not dominant, tool in supporting decision making; the proclamation of risk assessment/risk management as the dominant decision paradigm by Administrator William Ruckelshaus in 1983; a National Research Council recommendation that regulatory agencies should develop "inference guidelines" that would codify risk-assessment practices and reflect the latest scientific knowledge; and a growing body of scientific data on carcinogenesis that persuaded EPA to update its 1976 interim guidelines for cancer.[15]

In the spring of 1984 Deputy Administrator Alvin Alm charged EPA staff to develop risk-assessment guidelines for a number of health endpoints and also requested SAB review of the scientific adequacy of such guidelines. In the fall of 1984 EPA proposed five guidelines for cancer, chemical mixtures, developmental effects, exposure, and mutagenicity. The SAB review occurred in the spring of 1985.[16]

The SAB's role emerged in two phases. Within EPA it served as a scientific arbitrator among various EPA staff and their respective positions and between EPA and various public groups commenting on the proposed guidelines. Perhaps the most important aspects of the scientific debate in developing guidelines concerned two issues: How explicitly should EPA state the assumptions it used in preparing and using risk assessments? And how conservative should those assumptions be?

In general, the SAB supported retention of the conservative thrust of EPA's approach to risk assessment, but, to a significant degree, it encouraged EPA to state why such assumptions were necessary and how it would apply such assumptions, and it recommended that EPA encourage the submission of new information to enable periodic updating of the guidelines.[17] EPA largely adopted these recommendations.

The SAB review assumed a second role—that of legitimizer of the science-policy choices adopted in the guidelines—when EPA submitted the guidelines to the Office of Management and Budget (OMB) for review. OMB's staff, as well as selected staff from the Office of Science and Technology Policy and the National Science Foundation, raised a number of questions and objections relating to EPA's continuing use of conservative assumptions. Such objections, however, were of a philosophical nature rather than a substantive debate over scientific evidence. No new evidence surfaced during the OMB review.

As the OMB review process evolved over a series of months in the first half of 1986, individual congressmen began to express concern at the delay in finalizing the guidelines and over reports in the trade press that OMB staff were attempting to change their content.[18] A major factor in determining the outcome of the OMB review process was whether EPA enjoyed the support of the scientific community for the science-policy decisions embodied in the guidelines. The fact that scientists from academia, industry, public-interest groups, and other federal agencies had all served on the SAB review panels and had supported the positions adopted in the guidelines greatly legitimized EPA's position in the scientific community, on Capitol Hill, and, ultimately, within the federal government.

OMB completed its review of the risk-assessment guidelines in the latter part of the summer of 1986. EPA published five final guidelines—making no substantive scientific changes since the completion of the SAB review—on September 24, 1986.

1988 Report of the Research Strategies Committee: Reconciling Partnership and Criticism

For most of its history, the SAB has conceived of its role primarily in terms of responding to EPA requests for specific scientific reviews and as an advisory body as distinct from an approval body. This conception is consistent with congressional intent in creating the board. The congressional conference report leading to the legislative formation of the SAB states:

The [SAB] is intended to be advisory only. The Administrator will still have the responsibility for making the decisions required of him by law. The reviews and comments of the Board are to be provided to the Administrator for his use. The Board is not intended as a forum to be used by outside interests to criticize the workings of the Agency.[19]

Throughout its history the board has confronted two apparently contradictory tasks—to assist the administrator and EPA by reviewing the technical adequacy of information used for policy development, but to retain its independence from EPA. The board has attempted to reconcile these tasks by staying close enough to EPA to understand its needs but maintaining a sufficient distance not to compromise its own independence or be captured by EPA's internal bureaucratic competition. That board members themselves recognize this duality is perhaps the surest safeguard. Mitchell Small, a member of the SAB Environmental Engineering Committee, has stated, "We like the agency to feel comfortable with us and to come to us freely for help, but we have to guard ourselves to maintain an outside position."[20]

The board has not always succeeded in this task. During 1981–1982, as EPA's research budget was dramatically reduced, the board was a passive observer. It has been only moderately effective in alerting the agency to emerging issues. While the SAB's Ecology Committee attempted, in the mid- and late 1970s, to call attention to the acid deposition issue (without success), the board performed essentially no role in advising EPA of other issues whose importance was well known in the scientific community—issues such as the health risks from radon exposures, stratospheric ozone depletion, and global warming. EPA's inability to anticipate these and other issues stems, in part, from a Science Advisory Board that is overly reactive; in the words of SAB member Ellen Silbergeld, "We are like dance partners, waiting to be asked to dance."[21]

The 1988 report of the SAB's Research Strategies Committee reconciled the partner/critic dichotomy in as effective a manner as any activity in the SAB's history.[22] The committee sought to address a series of issues that, in the day-to-day management of various environmental problems, EPA and the SAB do not adequately address. These include the following:

• What is the mission of EPA?

• What is the role of research in a regulatory agency, and how is it best supported and managed?

• How can EPA obtain an earlier warning of emerging health and environmental problems?

Seeing itself both as partner and constructive critic of EPA, the committee recommended a number of major actions, including the following:

• EPA should shift the focus of its environmental protection strategy from end-of-pipe controls to preventing the generation of pollution. EPA should use a hierarchy of policy tools that support national efforts to (1) minimize the amount of wastes generated; (2) recycle or reuse the wastes that are generated; (3) control the wastes that cannot be recycled or reused; and (4) minimize human and environmental exposures to any remaining wastes.

• To support this new strategy, EPA should plan, implement, and sustain a long-term research program. In conjunction with EPA's program offices and the external scientific community, EPA's Office of Research and Development should develop basic core research programs in areas where it has unique responsibilities and capabilities.

• EPA must improve its capability to anticipate environmental problems. EPA should explicitly develop and use monitoring systems that help the agency anticipate future environmental conditions, and it should create a staff office that would be responsible for anticipating environmental problems and then recommending actions to address them.

- EPA should expand its efforts to assist all those parts of society that must act to prevent or reduce environmental risk. Since state, local, individual, and private-sector actions will become increasingly important for reducing the amount of waste and pollution generated, EPA needs to improve the education, training, technology transfer, and research programs that support such actions.

PREREQUISITES FOR AN EFFECTIVE SCIENTIFIC ADVISORY PROCESS IN REGULATORY AGENCIES

From the preceding examples, as well as other case studies in the literature, it is possible to state some prerequisites to maintaining effective and continuing independent scientific advice in the context of a regulatory agency. It is important to recognize, however, that no single form of peer review is adequate to address the range of issues and problems encountered by regulatory agencies in their development and use of scientific information. Therefore, different peer review processes, including the expanded use of advisory committees, have evolved to meet decision makers' needs for scientific quality.[23] These prerequisites include the following:

- Regulatory officials should believe that there is a positive incentive or, phrased another way, the absence of a negative incentive to invite the participation of external scientists and engineers into their decision process. These incentives may include the desire for a scientifically acceptable assessment of public health or environmental risks; or a concern about criticism if a scientifically inadequate document is used as a basis for decision making.

- In submitting a document for review by independent scientific committees, the regulatory agency staff needs to make explicit both the process and the logic by which it evaluated studies on the toxicity of a particular pollutant, calculated dose-response functions, integrated exposure data with the toxicity data, and arrived at a range of numbers that express the likelihood of the risk of a health or environmental effect. In other words, this "staff paper" should explicitly state the chain of scientific logic leading the staff to a particular scientific conclusion and policy recommendation(s).

- Independent peer review must be carried out early in the decision-making process. Two advantages of earlier involvement that are not present at later stages are (1) that it is easier to separate risk-assessment and risk-management issues and (2) that there is usually greater flexibility in addressing and resolving technical issues before an agency has formally proposed a particular regulatory action.

- Scientific advisors, besides having stature and authority within their professions, must deliver their advice in a timely manner and in a way that addresses the practical problems of the regulatory agency. For scientific advice to be used in the regulatory process, it should be transmitted in a form and according to a timetable that is compatible with the agency's needs. This assumes, of course, that scientific advisors receive appropriate lead time to review technical documents and prepare scientific reports.

- Scientific advisors need to know if their advice will or will not be taken. Most scientists do not expect that their advice will be completely accepted, but they strongly desire that the regulatory agency inform them of the degree to which it will use such advice, and if not, why not.

- Scientists should interact with both the staff and senior managers of an agency on a frequent basis, and not only at formally scheduled public meetings. There is a need for frequent and less formal exchanges that can serve to clarify the objectives and operating methods of peer reviewers while building personal trust between the advisors and the agency and avoiding surprises. Agency officials should not be surprised at the conclusions of their advisors and, in addition, should have the opportunity to discuss the advisors' findings before a final report is issued. In the business of providing scientific advice, familiarity breeds trust and understanding and need not jeopardize independence.

- There must be continuity in the membership of advisory committees. This is necessary to develop a sense of institutional memory between the advisors and the agency and assure the accuracy of scientific advice. Continuity also promotes more predictable and efficient committee review procedures, induces a common sense of mission among committee members, and enhances the likelihood that a regulatory agency will give more serious consideration to an advisory report, if only because the advisory relationship is a continuing one. Continuity of membership does not remove the need for a routine process of rotating scientists and engineers on and off committees on a periodic basis to introduce new scientific views and perspectives.

- Scientific advisory committees or boards should adopt explicit guidelines to protect themselves from conflict of interest or the appearance of conflict of interest. Such guidelines can enhance both the integrity and the authority of the advisory process.

- The scientific advisory process must be a public process. This is necessary not only to comply with certain legal requirements of the Federal Advisory Committee Act but also to ensure the credibility of the scientific review process. A public advisory process, allowing some form of public participation, can yield several important benefits. It can lead to the introduction of new and important scientific information by members of the public; it enables the regulatory agency to identify public concerns before it issues a formal proposed regulation; and it can lead to consensus on key scientific issues in a manner that is more acceptable to the public because of the openness of the advisory proceedings.[24]

These prerequisites do not constitute absolute requirements for an effective scientific advisory process. But in the experience of the Science Advisory Board, they have proven to be reliable and durable indicators for guiding and evaluating the board's performance and its working relationship with EPA.

POTENTIAL NEW DIRECTIONS AND CHALLENGES FOR EPA AND THE SAB

America is in the midst of a new environmental activism to ameliorate real and perceived health and environmental risks. This activism has taken and will continue to take the form of increased support for government regulation of business activity, enactment of additional legislation to respond to existing and newly recognized environmental issues, and increased pressure upon EPA to more rapidly respond to regional, national, and global problems.

What are the implications of this activism for EPA and the Science Advisory Board? There are at least six possibilities:

- During the early years of its history, scientists did not perform a major role in EPA decision making because of the perceived need to organize crisis management responses to imminent environmental problems. Legislation for hazardous waste management and newly enacted Clean Air Act Amendments are premised upon the existence of an environmental crisis and posit a reduced role for the scientific community in solving environmental problems and ensuring the technical adequacy of EPA decision making. Science will perform a reduced role as a legitimizer of EPA policies for specific pollutants and industrial processes.

- EPA will experience increased pressure to modify its risk-assessment/risk-management framework for decision making. This framework is predicated upon the role of science as a legitimizer of policy choices. Instead, EPA will, over time, adopt measures and strategies aimed at pollution prevention and reduction. Risk reduction, not risk management, will emerge as the major decision framework for the 1990s.

- EPA will focus less on chemical-by-chemical regulations and earmark more resources to reducing exposures across a number of industrial and consumer activities and classes of pollutants.

- The SAB should conduct fewer reviews of individual compounds and place more emphasis on advising EPA how to better identify emerging environmental problems; accelerate the transfer of information and technology from the laboratory to the field; leverage EPA research programs with those of other federal agencies and the private sector; integrate interdisciplinary environmental problem solving into university curricula; and maintain the infrastructure of people and equipment devoted to environmental problems.

- The SAB should devote more emphasis to identifying advances in managing research and in technical information, urging EPA to more readily incorporate advances in the computer and social sciences to manage information and people.

- Using available technical information and the professional judgment of its members, the SAB should also see its role as one of advising EPA on health and environmental priorities on a continuing basis.

NOTES

1. 42 U.S.C. 4365(a)(c)(e).
2. Ibid.
3. U.S. Environmental Protection Agency (EPA), Science Advisory Board, *Report of the Director of the Science Advisory Board for Fiscal Year 1987*, SAB–88–007, December 1987, p. 10.
4. For a discussion of the CHESS program and controversy, see U.S. EPA, *Health Consequences of Sulfur Oxides: A Report from CHESS, 1970–1971*, EPA–650/1–74–004, May 1974; U.S. EPA, Science Advisory Board, *Review of the CHESS Program: A Report of a Review Panel of the Science Advisory Board*, March 14, 1975; U.S. House

of Representatives, *The Environmental Protection Agency's Research Program with Primary Emphasis on the Community Health and Environmental Surveillance System (CHESS): An Investigative Report*, November 19, 1976, Washington, D.C., GPO, 1976 (also known as the Brown report); U.S. EPA, Science Advisory Board, *Report of the Health Effects Research Review Group*, February 1979, pp. 55–60; W. W. Holland et al., "Health Effects of Particulate Pollution: Reappraising the Evidence," *American Journal of Epidemiology*, vol. 110, November 1979, pp. 525–659.

5. Brown report.

6. For a discussion of EPA's development of the 1979 ozone standard, see Christopher H. Marraro, "Revising the Ozone Standard," in *Quantitative Risk Assessment in Regulation*, edited by Lester B. Lave, Brookings Institution, Washington, D.C., 1982, pp. 55–97; Lawrence J. White, *Reforming Regulation: Processes and Problems*, Prentice-Hall, Englewood Cliffs, N.J., 1981, pp. 47–70.

7. American Petroleum Institute v. Douglas M. Costle, Administrator, Environmental Protection Agency, 661 F. 2d 340 (D.C. Cir. 1981), cert. denied.

8. Statement of Stephen J. Gage, Assistant Administrator for Research and Development, U.S. Environmental Protection Agency, before the Subcommittee on Health and the Environment, Committee on Interstate and Foreign Commerce, U.S. House of Representatives, November 27, 1979; American Petroleum Institute, *The Statutory Authority and Responsibilities of the Clean Air Scientific Advisory Committee*, June 7, 1979; memorandum from Lester D. Grant, Joseph Padgett, and Terry F. Yosie, "Recommended Procedures for Involving the Clean Air Scientific Advisory Committee in the Review Process for National Ambient Air Quality Standards," U.S. EPA, June 14, 1979; U.S. EPA, Science Advisory Board, *Setting Ambient Air Quality Standards: Improving the Process*, September 1981; Michael A. Berry, *A Method for Examining Policy Implementation: A Study of Decision-Making for the National Ambient Air Quality Standards, 1964–1984*, U.S. EPA, Office of Research and Development, EPA–600/X–84–091, May 1984.

9. Memorandum from Anne M. Gorsuch to Assistant Administrators, "Improving the Scientific Adequacy of Agency Regulations and Standards," February 8, 1982.

10. 42 U.S.C. 7417(f)(4)(1982).

11. Memorandum from William D. Ruckelshaus, Administrator, to Chairman, Science Advisory Board, December 6, 1983.

12. U.S. EPA, Science Advisory Board, *Report on the Scientific Basis of EPA's Proposed National Emission Standards for Hazardous Air Pollutants for Radionuclides*, August 1984.

13. Ibid.

14. Memorandum from Lee M. Thomas, Administrator, to Assistant Administrators, "Improving the Agency's Use of the Science Advisory Board," June 25, 1985.

15. National Academy of Sciences/National Research Council, *Risk Assessment in the Federal Government: Managing the Process*, National Academy Press, Washington, D.C.: 1983; Terry F. Yosie, "Science and Sociology: The Transition to a Post-conservative Risk Assessment Era," in James J. Bonin and Donald E. Stevenson, eds., *Risk Assessment in Setting National Priorities*, Plenum Press, New York, 1989, pp.1–11.

16. Memorandum from Alvin L. Alm, Deputy Administrator, to Assistant Administrators, May 21, 1984.

17. Yosie, "Science and Sociology," p. 5.

18. Letter from John D. Dingell to Lee M. Thomas, July 17, 1986.

19. H. R. Conf. Rep. no. 722, 95th Cong., 1st sess., 1977, p. 16, 1977 U.S. Code Cong. and Admin. News, p. 3295.

20. Mitchell Small, quoted in Susan R. Jones, "Scientists Take on a Bigger Role in EPA Decisions," *Chemical Week*, January 28, 1987.

21. Ibid.

22. U.S. EPA, Science Advisory Board, *Future Risk: Research Strategies for the 1990s*, SAB-EC–040, September 1988.

23. Terry F. Yosie and Janis C. Kurtz, "Editorial: Peer-Review Processes Used in Regulatory Decision Making," *Environmental Toxicology and Chemistry*, vol. 6, 1987, pp. 491–93.

24. U.S. EPA, Science Advisory Board, *Report of the Director of the Science Advisory Board for Fiscal Year 1986*, SAB–87–007, October 1986, pp. 3–5.

REFERENCES

Case Studies of SAB Reviews

Lippmann, Morton. "Role of Science Advisory Groups in Establishing Standards for Ambient Air Pollutants." *Aerosol Science and Technology*, vol. 6, 1987, pp. 93–114.

EPA Risk Assessment/SAB Role

North, D. Warner, and Terry F. Yosie. "Risk Assessment: What It Is, How It Works." *EPA Journal*, vol. 13, November 1987, pp. 13–15.

Yosie, Terry F. "Some New Directions for the Science Advisory Board and EPA Science Policy." Paper presented before the American Industrial Health Council, February 25, 1986.

———. "EPA's Risk Assessment Culture." *Environmental Science and Technology*, vol. 21, June 1987, pp. 526–31.

———. "Science and Sociology: The Transition to a Post-conservative Risk Assessment Era." In James J. Bonin and Donald E. Stevenson, eds. *Risk Assessment in Setting National Priorities*, Plenum Press, New York, 1989, pp. 1–11.

Peer Review in Regulatory Agencies

Burack, Thomas S. "Of Reliable Science: Scientific Peer Review, Federal Regulatory Agencies, and the Courts." *Virginia Journal of Natural Resources Law*, vol. 7, Fall 1987, pp. 27–110.

Jasanoff, Sheila. "Peer Review in the Regulatory Process." *Science, Technology, and Human Values*, vol. 10, Summer 1985.

Yosie, Terry F., and Janis C. Kurtz. "Editorial: Peer-Review Processes Used in Regulatory Decision Making." *Environmental Toxicology and Chemistry*, vol. 6, (1987), pp. 491–93.

American Chemical Society. "Issues in Peer Review of the Scientific Basis for Regulatory Decisions." 1985.

CHAPTER 3

THE CHEMICAL INDUSTRY INSTITUTE OF TOXICOLOGY

Robert A. Neal

The publication of *Silent Spring* by Rachel Carson in 1962 marked the beginning of an increased societal concern about the contamination of the environment and the workplace with toxic chemicals. These concerns led to the passage of the Clean Air Act, the Clean Water Act, and the National Environmental Policy Act. Initially, the chemical industry was relatively untouched by these legislative initiatives. However, in 1974 it was discovered that several workers at a plant in Kentucky, while working with vinyl chloride, a precursor of the plastic polyvinyl chloride, developed cancer of the liver. This unsettling discovery had a very profound effect on the chemical industry. Were there other chemicals that were causing similar effects in exposed workers? The challenge was to determine which chemicals were dangerous and under what conditions, and how these could be handled to ensure a proper degree of safety in their use. To answer these kinds of questions, the Chemical Industry Institute of Toxicology (CIIT) came into being.

ORIGINS OF CIIT

The primary credit for the concept of CIIT must go to the Dow Chemical Company. More correctly, the credit should go to Mac Pruitt, who was then the vice president for research and development at Dow. Basically, the idea was a simple one, that industry should take the offensive, pool its resources, and get sound scientific information on the potential health effects of chemicals used widely throughout the chemical industry. The first of a series of meetings that eventually led to the organization of CIIT was held in August 1974. Seven companies were represented at that first meeting: the Dow Chemical Company, E. I. du Pont de Nemours & Co., Inc., Monsanto Company, Stauffer Chemical

Company, Exxon Chemical Company, Union Carbide Corporation, and Air Products and Chemicals, Inc. A second meeting was held shortly after, and the original seven were joined by four more companies: Diamond Shamrock, Eastman Kodak, Allied Corporation, and Shell Chemical Company.

In December 1974 a certificate of incorporation was filed in Delaware. By April 1975 a working group of the member companies had produced a document that formed the blueprint for CIIT. This blueprint has changed little with time. This document recognized the problems that were resulting from an increasingly complex technologically based society and expressed the need for a coordinated effort to understand and solve the problems. The document also recognized the chemical industry's responsibility and right to provide leadership in the area. CIIT's objective was to provide a sound scientific industry presence in chemical safety evaluation. CIIT was to generate data about compounds used broadly throughout the chemical industry and to use this information to assess the risk of exposure of humans to these chemicals. CIIT also had as a goal the improvement of the methodology used in assessing the potential toxicity of chemicals. Finally, CIIT had the obligation to disseminate information about the potential toxicity of chemicals widely and quickly and to promote the training of toxicologists and other scientists who were at that time in short supply.

In mid–1975 the founding members were still uncertain whether a separate laboratory was needed or whether the work planned could be contracted out to existing laboratories. Considerable time was spent on this question. Finally, the decision was made that CIIT should have its own facilities and should be staffed with scientists from the various disciplines important in assessing the potential toxicity of chemicals. It was also decided at that time that two-thirds of the funds of the institute would be spent on testing of chemicals for their potential to cause toxic effects in humans, whereas the other one-third of the resources would be applied to basic research in toxicology.

The remainder of 1975 and 1976 were spent in recruiting new members to CIIT, selecting the site for the new laboratory, and deciding who would be the first president of CIIT. By September of 1975 six new companies had joined the original eleven. A site selection committee was also active, looking at places like Houston, St. Louis, Oak Ridge, and Research Triangle Park. Research Triangle Park in North Carolina was finally selected as the site for the new laboratory. Among the factors that led to the selection of Research Triangle Park was the near proximity to a number of excellent universities and medical facilities. It was also close to Washington, D.C., if the scientific staff of CIIT were needed there. Several government laboratories, such as EPA and National Institute of Environmental Health Science (NIEHS) were already there, and land was available within the park for purchase. After a long search, Leon Golberg was selected as the first president. He was hired in February 1976. By the end of 1976, fourteen additional people had been employed. Temporary quarters were leased in Raleigh, North Carolina, in August 1976, and the research activities began.

The period from late 1976 to 1979 was spent in expanding the membership in CIIT and in overseeing the building of the current facilities. A contract was let to build the laboratory for occupancy within twelve months starting in September 1977. However, strikes, weather, and other factors, including a very complex facility, delayed the completion of the project. The building was not ready for occupancy until 1979, and it was not completely operational until late 1980.

While the new facilities were being built, the research program was under way at temporary facilities in Raleigh, North Carolina. The program consisted of both basic research in mechanisms of toxicity and letting contracts to outside laboratories for testing of widely used commodity chemicals for their potential toxicity. CIIT personnel growth continued. Thirty-eight were employed by the end of 1977. This increased to sixty-seven in 1978, and eighty-one in 1979. The expenditures grew from $2.5 million in 1977 to $4.2 million in 1978 and $6.7 million in 1979.

Up until late 1979 few people were aware of CIIT outside of the toxicological community and the chemical industry. However, in October 1979 CIIT announced that in an interim sacrifice of rats exposed to formaldehyde by inhalation there was strong evidence for the compound-related production of nasal cancer in these animals. Since formaldehyde is a high-volume chemical with numerous applications, the news that it was carcinogenic in rats focused a lot of attention on CIIT.

By 1980 CIIT had grown to 92 permanent staff and a budget of $8.3 million. The size of the institute has continued to grow, and in 1988 there were 119 permanent employees with an annual budget (1988) of $13.3 million. This excludes approximately 30 postdoctoral or visiting scientists who are usually in residence at CIIT at any one time.

Since 1980 the mission of CIIT has also undergone a slow change. When CIIT was initially founded, two-thirds of the resources of the institute were devoted to the testing of chemicals for their potential toxicity and one-third to more basic studies. Over the past eight years the routine testing of chemicals for their potential toxicity has been gradually phased out. Currently, the entire resources of the institute are devoted to examining the mechanisms of toxicity of environmentally important industrial chemicals, examining the relevance of animal tests to humans, and developing new methodologies for determining the potential of toxicity of chemicals to humans.

ORGANIZATION

The Chemical Industry Institute of Toxicology is a not-for-profit scientific research organization. CIIT is neither owned by the members nor devoted to their exclusive interests. CIIT has characteristics of a joint research effort in which its members are banded together to ensure that mutually significant scientific research is pursued, but it is different in that the interests involved are shared by the membership with the public and industry at large. CIIT also has

Table 3.1
CIIT Supporting Companies

Air Products and Chemicals, Inc.	Lyondell Petrochemical Co.
Allied Signal, Inc.*	Mallinckrodt Inc.
Amoco Chemical Company	Manville Corporation
ARCO Chemical Company	Mobil Research & Development Corporation
Bayer USA Inc.	Monsanto Company
BFGoodrich	Nalco Chemical Company
BP America, Inc.	National Starch and Chemical Company
Chevron	NOVA Corporation of Alberta
CIBA-GEIGY Corp.	Occidental Chemical Corporation
The Dow Chemical Company*	Olin Corporation
Dow Corning Corporation	Phillips 66 Company
E.I. duPont de Nemours & Co., Inc.*	PPG Industries, Inc.
Eastman Kodak Company*	The Proctor & Gamble Company
Ethyl Corporation	Quantum Chemical Corporation
Exxon Chemical Company*	Reichhold Chemicals Inc.
Fina Oil and Chemical Company	Rhône-Poulenc Inc.
FMC Corporation	Rohm and Haas Company
GAF Corporation	Sandoz Crop Protection Corporation
General Electric Company	Shell Chemical Company*
Georgia Gulf Corporation	Texaco Inc.
W.R. Grace & Co.	Union Carbide Corporation*
Hoechst Celanese Corporation	Unocal Corporation
Hüls America Inc.	Volkswagon AG
ICI Americas, Inc.	Vulcan Materials Co.
The Lubrizol Company	

* Founding members

some characteristics of a trade association. However, CIIT is different from traditional associations because it achieves member-company benefit through scientific pursuit and public dissemination of information of significant value to federal and local governments, the public at large, other industries, and the chemical industry, not by lobbying or advocating parochial views, activities that are typical of trade associations. CIIT has many of the characteristics of an educational or scientific institution, such as a university or public research foundation, because it is devoted to scientific endeavor and the open publication of information it develops. But it is also different because it is dedicated to pursuing these goals in the interest of the public and its member companies and not strictly for educational or scientific purposes.

Table 3.1 is a listing of the supporting companies of CIIT. The founding members are indicated by an asterisk.

The affairs of CIIT are managed by its Board of Directors. The Board of Directors is composed of representatives from the member companies. In addition, there are five members from outside the company membership. The outside members come largely from academic institutions. Internally, the institute is managed by the president, who in turn reports to the Board of Directors. He is assisted by a director of administration, who has responsibility for support functions, such as building services, accounting, purchasing, laboratory compliance, and information sciences.

The research program of CIIT is organized by departments. These are the Departments of Experimental Pathology and Toxicology, Biochemical Toxicology, Risk Assessment, Cellular and Molecular Toxicology, and Epidemiology.

The Department of Experimental Pathology and Toxicology has two major objectives. The first is to utilize the available expertise in animal pathobiology and inhalation toxicology to address important and timely topics on the mechanisms of toxicity of commodity chemicals. Second, the laboratory animal care group, the inhalation toxicology laboratories, and the pathology sections of histology, electron microscopy, and clinical pathology also provide appropriate services to support the research programs of this and other departments.

The Department of Biochemical Toxicology and Pathobiology carries out studies of the metabolism and pharmacokinetics of chemicals of interest to CIIT. Particular attention is given to mechanisms of toxic action involving the metabolic formation of transient or persistent toxic biotransformation products. The study of exogenous protective systems and the biological responses when two or more chemical exposures are combined are also of concern. The work in this department and its collaboration with other departments of the institute are in part facilitated by the presence of a strong analytical component. Methods for determining dose-response relationships between environmental exposure and internal target-site dosimetry are being developed for use in quantitative risk assessment.

The Department of Cellular and Molecular Toxicology's major research emphasis is mechanistic toxicology. This effort utilizes both whole-animal and in vitro exposure models and provides systems to study mechanisms of toxic injury at the cellular and molecular level. Through an understanding of the mechanisms underlying toxic manifestations of chemical exposure, advances can be achieved that should improve risk assessment and reduce animal usage.

The Department of Epidemiology strives to provide a link between laboratory studies on animal models (and other systems) on the one hand and the practical realities of occupational exposure on the other. A primary function of this department is to provide leadership and organization for the utilization of information records and other resources available in industry relating to the effects or lack thereof of occupational exposure to selected compounds of interest to CIIT. The institute's laboratory studies often suggest approaches to monitoring of effects in the workplace, and conversely, the provision of data and specimens from occupationally exposed personnel are very valuable to the experimental programs.

The responsibility of the Department of Risk Assessment is to help CIIT more effectively accomplish its primary mission: generation of research results that will provide an objective and scientific basis for assessing human health effects from chemical exposure. Particular attention is given to the problem of incorporating information regarding mechanisms of toxic action in the mathematical models that are utilized for quantitative risk estimation. The program seeks ways to better utilize existing scientific information in the risk-assessment process as it is presently structured.

The Postdoctoral Fellow Program consists of approximately 30 qualified scientists who receive advanced research training in the CIIT laboratories. Each fellow normally spends two years at CIIT before seeking permanent employment in other toxicology research facilities. Postdoctoral fellows are selected from applicants recently completing doctoral degrees in biological or physical sciences related to toxicology. The goal of the postdoctoral training program is to provide in-depth training in a selected area of toxicology as well as exposure to several areas of toxicology.

Under certain special circumstances it is desirable for CIIT to temporarily bring to the institute a senior investigator with training beyond the doctoral level. Consequently, CIIT has a limited visiting scientist program. The recipient of this award may be an individual with unique skills necessary for a special research project, a scientist from an institution or country who would add to the status of CIIT as a leader in toxicology, a CIIT postdoctoral fellow who has completed his/her two-year program and has been invited to complete a project of special interest or value to the institute, or a physician who may participate in a clinical aspect of CIIT's research. Typically, a visiting scientist would participate in the research program of the institute for approximately one year, with the area of research previously agreed to by the visiting scientist and the president of CIIT.

OPERATING POLICIES

There are five basic policies governing CIIT's operations. The first of these is that CIIT member companies will not gain competitive advantage over nonmembers by virtue of their membership in CIIT. CIIT's research efforts will inevitably affect members and nonmembers, some favorably and some unfavorably. The policy requires that the chips fall where they may so that any scientific impact will be strictly the result of the evidence and totally unrelated to CIIT membership.

Second, research results generated by the institute are published or otherwise made public when evaluated and ready for disclosure and are made equally available to all interested parties. CIIT member companies are not favored by advance disclosure of research results. The president of CIIT and the scientific staff have sole control over the content and timing of the release of information generated at the institute.

A third fundamental policy of CIIT is that it will conduct its activities in a manner designed to prevent harm to third parties. Obviously, significant institute research disclosures may adversely affect the commercial lives of third parties as well as members. These disclosures, however, should not per se be construed as unreasonable activities.

Fourth, the institute conducts its operations in compliance with all legal requirements and in accordance with the highest ethical standards. In this regard, CIIT policy prohibits its personnel from engaging in lobbying activities. It particularly prohibits the influencing of governmental action by any means other than the disclosure and interpretation of scientific data.

Finally, when the president of CIIT determines that it is in the public interest and CIIT's interest to do so, CIIT may participate voluntarily in adversarial proceedings. CIIT will cooperate, appear, and testify without compensation or reimbursement of expenses. CIIT participation in adversarial proceedings in which it is not a party, whether voluntary or by compulsion, will be nonadversarial. CIIT will satisfy itself that all parties have appropriate and timely access to the information furnished by it to any party or to it by any party. Its response to requests will not vary predicated upon the adversarial position of the requesting party. CIIT's participation in these adversarial proceedings will be limited to expert testimony on scientific matters related to its respective fields of expertise.

SELECTION AND OVERSIGHT OF THE RESEARCH PROGRAM

Suggestions for research projects are solicited each year from member companies. Each year a subcommittee of the Board of Directors, the Development Committee, meets with the scientific staff of CIIT to examine the suggested list of new research projects and evaluate these suggestions in light of the research underway at CIIT at that time. The purpose of this meeting is to examine whether the research program of the institute is being responsive to what is perceived by the member companies to be the most pressing research needs. The Development Committee acts in an advisory capacity. Which research projects will be carried out at CIIT is a decision made by the president of CIIT and the scientific staff.

The scientific quality of the research program and the scientific staff at CIIT is reviewed each year by an outside group of scientists. This Scientific Advisory Panel usually numbers fifteen to twenty members drawn from academia, industry, and government laboratories. The members of the panel are chosen based on their expertise in the areas of research underway at CIIT at the time of the review. The quality of each individual project and the qualifications of each investigator are evaluated in written critiques. At the same time, the Development Committee evaluates the relevance of the research projects to research goals of CIIT.

These yearly reviews help ensure that a high-quality and relevant research program is being conducted at CIIT. In addition, the Scientific Advisory Panel and the Development Committee provide helpful suggestions for changes in approach or new approaches to the problems under investigation at CIIT. The findings of the Scientific Advisory Panel and the Development Committee are provided verbally and in writing to the Board of Directors.

RESEARCH PROGRAM

The ultimate objective of the CIIT research program is to provide a scientific basis for estimating the risk to humans from incidental, intentional, or occupational chemical exposures. In order to accomplish this objective, the research program of CIIT is designed to achieve a number of technical goals. These goals are the following:

1. Improving the scientific basis for human risk assessment through the use of comparative absorption, disposition, pharmacokinetic, and target-organ data from laboratory animals and humans—with special emphasis on cancer risk.
2. Elucidating mechanisms of toxic action of selected chemicals—then developing methodology that facilitates timely expansion of key findings from this research to studies on a broad range of compounds (e.g., extending formaldehyde approach and findings across many more chemicals).
3. Development of biological markers that can detect early significant toxic effects in animal models and humans, assist in identification of sensitive individuals, and/or improve the prediction of safe levels in humans.
4. Development of improved methods—both in vitro and in vivo—to supplement or replace current animal assays for assessing human risk.
5. Development of improved methods for understanding of potential health risks from inhalation of fine particles or fibers.
6. Understanding the relevance of the results of animal tests—such as rodent liver tumors, rat kidney toxicity or carcinogenicity, and mouse lymphoma—in the assessment of human cancer risk.
7. Development and validation of methods for evaluating health effects resulting from chemicals that may impair reproduction, from chemicals that may be neurotoxic or immunotoxic, and from recombinant DNA organisms or products.
8. Improving the understanding of potential chronic health effects from short-term or intermittent high-level exposure to a chemical compared to a low-level exposure over an extended time period.
9. Improving the understanding of potential health effects from exposure to mixtures of chemicals versus exposure to a single chemical.

This list of technical goals is not static. It is reexamined periodically to determine if it requires revision in order to better reflect the research priorities of the scientific staff of CIIT and its Board of Directors. Table 3.2 gives a representative list of research projects at CIIT.

SOME HIGHLIGHTS OF THE RESEARCH PROGRAM

One of the major accomplishments of the CIIT research program is the demonstration of the utility and validity of the use of ''target-site'' dose rather than exposure dose in calculating dose-response relationships at low levels of expo-

Table 3.2
CIIT Research Projects

Project Number	Title
1224	Mechanisms Involved in Teratogenesis
1230	Biology of Hepatocellular Preneoplastic Lesions
1235	Model Systems for the Study of Respiratory Epithelial Toxicity
1240	Mechanisms of Chemically-Induced Leukemia/Lymphoma
1243	Mechanisms of Benzene-Induced Leukemogenesis
1253	Mechanisms of Hydrocarbon-mediated Nephrotoxicity
1271	Mechanisms of Acrylonitrile Toxicity and Carcinogencity
1276	Biochemical Toxicity of Aldehydes
1277	Formaldehyde Toxicity in Monkeys; Pathology, Cell Turnover, Covalent Binding and Risk Assessment
1281	Mechanisms of Dioxin Toxicity to Human Keratinocytes
1312	Sequence Specificity of Mutagenesis in Bacteria and Human Cells
1313	Development of a Transgenic Mouse for Mutagenicity Studies
1350	Mammalian Cell Transformation and Promotion Assays
1360	Chemically Induced DNA Damage, DNA Repair and Hyperplasia
1365	Nongenotoxic Mechanisms in Carcinogenesis
1374	Reproductive Surveillance of Chemical Industry Workers
1378	Mortality Study of Automotive Foundry Workers Exposed to Formaldehyde and Other Agents
1381	Cell-Specific Factors in Carcinogenesis and Toxicity
1382	Nasal Function and Pathogenesis of Chemically-Induced Nasal Toxicity

sure. These studies have shown that the target-site dose is not necessarily linearly related to the exposure dose. In the case of formaldehyde, the target-site (nasal epithelial cell DNA) dose at low exposure doses (0.3, 1.0, 2.0 ppm) is less than would be predicted by the target-site dose at higher exposure doses (6, 10, 15 ppm).

A number of compounds (gasoline, 1,4-dichlorobenzene, limonene, jet fuels, isophorone, and so on) produce kidney tumors in male (but not female) rats in chronic studies. CIIT's investigations of this effect of these compounds have shown that it likely results from the compound-induced accumulation of alpha-2-microglobulin in the P–2 segment of the kidney tubules. If the accumulation of alpha-2-microglobulin is sufficient, cell death results, and there is a rapid turnover of cells in this segment of the kidney. Additional studies have shown that the chronic cell turnover likely acts as a promotional stimulus to "spontaneously" initiated cells in the male rat kidney. What appears to be required for these tumors to develop is, first, the excretion of large amounts of a proteolytic-enzyme-resistant, low-molecular-weight protein by way of the kidney, and second, that exposure to the compounds listed here increases the resistance of this low-molecular-weight protein to proteolytic enzymes. Neither female rats nor humans, in contrast to male rats, normally excrete large amounts of this protein in the urine. This explains why female rats do not develop kidney tumors on exposure to these compounds. In addition, it is unlikely that humans will develop tumors on exposure to these compounds, at least by the mechanism determined for male rats.

Research at CIIT has led to the development of an in vivo test for genotoxicity that has been widely adopted by the scientific community. It is called the in vivo–in vitro test for genotoxicity. The test consists of exposing rats to the compound of interest followed by isolation of cells from the tissues of interest (liver, kidney, bronchial epithelium, nasal respiratory epithelium) and exposing these cells to a pulse of tritiated thymidine followed by layering on a photographic emulsion. After a suitable exposure time, the emulsion is developed, the radioactivity over the nucleus as compared to that over the cytosol is determined, and the results are compared with those obtained with control animals. In this way the ability of specific chemicals to cause damage to the DNA of the cells of the organs of interest can be determined.

Other areas where CIIT's research has resulted in important progress are studies of the mechanism of tumor formation in the livers of rodents exposed to peroxisome proliferating agents, studies of the mechanisms of carcinogenicity of the dinitrotoluenes, an improved understanding of the role of virus in chemically induced lymphoma in mice, and a better understanding of specificity of alkylating agents in inducing mutations in bacterial and human DNA.

In sum, CIIT is a not-for-profit research institute devoted to improving our ability to determine the potential for chemicals from various sources to cause toxicity in exposed human populations. The major source of the funds for the research programs of the institute are dues paid on a quarterly basis by its member companies. The nature and stability of the funding and the independence of the institute from its funding source have allowed it to focus on long-term interdisciplinary efforts leading to a better understanding of some of the most important issues in toxicology.

CHAPTER 4

THE HEALTH EFFECTS INSTITUTE

Thomas P. Grumbly

Two persistent themes in the politics of regulatory reform have been the importance of improving the science base underlying regulation and the advisability of reducing the adversarial quality of the debate between the public and private sectors. The quest for "sound science" is so universal that the words themselves have almost taken on the quality of a chant, and some people regard them as code words for less regulation. The desire for a less adversarial, more civil quality to our debates is one not universally held, particularly by those who are trained in the adversarial process and who believe that the achievement of just outcomes inevitably requires crossed swords in American society. Both the persistence of the themes and the presence of doubters justify an examination of an institution explicitly devoted to both improved science and less adversarial discourse. That institution is the Health Effects Institute (HEI), established in late 1980 by the Environmental Protection Agency and the automotive industry to be the nation's center for research into the health effects of automotive emissions.

The HEI is a nonprofit corporation based in Cambridge, Massachusetts. It is independent of both the government and the industry. It is governed by a three-member Board of Directors and chaired by former Watergate Special Prosecutor Archibald Cox. It is jointly and equally funded by the federal government and private industry to the tune of about $6 million per year. Most of its work is currently undertaken by scientists working on nearly fifty projects at forty universities in the United States, West Germany, the United Kingdom, and Canada. In addition to the Board of Directors, the institute has two separate technical committees. The Health Research Committee plans and commissions research based upon the requests of the sponsors but ultimately reflecting its own individual judgments about the importance of problems and the likelihood of scientific progress. The Health Review Committee, which has no role in the selection or oversight of research, directs a post hoc peer review process of investigators' research and develops a criticism of the institute's sponsored work that accompanies the research into publication. The EPA has agreed to use HEI's work directly and not to subject it to any further peer review process.

It can fairly be said that HEI is unique in its independence, its mode of financing, its utilization of scientific committees to plan and criticize its own work, and its fundamental goal: to provide a commonly accepted scientific base upon which all interests (governmental, industrial, academic, and environmental) can rely. It does not propose to eliminate conflict; rather, it supposes that by removing science from the adversarial process, one can both improve the science and clarify the basis of conflict. This chapter examines both the validity of the concept and the quality of the implementation. It does so by

- briefly examining the public policy context within which HEI was created to see what specific elements made it possible;
- discussing the structure and process of the institute to see how the founders of HEI attempted to tailor it to meet a set of perceived problems;
- examining the implementation of the institute to date to see how well the institute is meeting its own objectives and dealing with the problems that were originally perceived; and
- evaluating, on a preliminary basis, the validity of the concepts upon which the Institute is based.

THE CLEAN AIR ACT

The federal Clean Air Act is the national air pollution control law.[1] It was first passed in 1963 and significantly revised in 1967, 1970, 1974, 1977, and 1990. The essential purpose of the law is simple: to protect the public health and the environment from man-made air pollution.

Three main elements comprise the regulatory tools of the Clean Air Act:

- National ambient air quality standards (NAAQSs) and deadlines are targets set by EPA to define how clean the air should be to protect the public health, as well as a set of deadlines for achieving these levels.
- Federal new motor vehicle emission standards are a set of enforceable national limits on the amount of pollution that may be emitted from new cars and trucks.
- State implementation plans (SIPs) are a set of enforceable emission limitations for chemical companies, power plants, and other stationary and moving sources of air pollution. Each state's plan must be sufficient to achieve the required level of clean air set in the NAAQS by the deadlines established in the law.

EPA revises its NAAQSs through the use of "criteria" documents, which compile and analyze the studies that have been done on the health and environmental effects associated with various air pollutants. These criteria documents are subject to review by EPA's Science Advisory Board and to public comment before they become final. There are now criteria documents and NAAQSs on six pollutants: sulfur oxides, particulate matter, carbon monoxide (CO), ozone, nitrogen dioxide, and lead. Four of these (CO, ozone, nitrogen dioxide, and particulates) are particularly relevant to motor vehicles. EPA is required to review these standards every five years.

In the 1970 Clean Air Act Amendments Congress decided that air pollution from cars was not being controlled well enough. Consequently, it took the very unusual step of actually specifying in the law that emissions of the major automotive pollutants had to be reduced by 75-90%, effective for all new cars built after 1976. Although some delays have subsequently been granted, these congressionally established emission limitations for new cars remain applicable today. The auto pollutants covered by these standards are carbon monoxide (CO), hydrocarbons (HC), and oxides of nitrogen. The latter two of these react in the presence of sunlight to produce ozone, one of the most difficult of the NAAQSs to attain.

THE 1977 CLEAN AIR ACT AMENDMENTS

J. Ronald Fox has pointed out that "the 1977 amendments to the Clean Air Act heightened the animosity between the EPA and the automobile industry."[2] Specifically, the amendments "increased the responsibility of both manufacturers and the EPA to generate and evaluate additional health data on auto emissions."[3] Section 202(a)(4)(A) mandated that the administrator of EPA ban any devices in new vehicles after 1978 that contribute to an unreasonable risk to the public health ("Effective with respect to vehicles and engines manufactured after model year 1978, no emission control device, system, or element of design shall be used in a new motor vehicle or new motor vehicle engine if such device, system, or element of design will cause or contribute to an unreasonable risk to public health, welfare, or safety in its operation or function").[4] Additional heavy-duty provisions were also passed to provide the EPA administrator with a process to revise emission standards for heavy-duty vehicles without changing the statute.[5]

By the time Congress passed these amendments, the environment surrounding the political health effects of air pollutants was itself poisonous. Industry believed, and with some justification, according to former Administrator Douglas Costle, that EPA's initial research on automotive emissions was "fairly mediocre."[6] From the agency's perspective, there was a fear about the objectivity of manufacturers' studies, as well as a general belief that manufacturers would overwhelm the agency with duplicative studies.[7] As Costle put it, the implementation of Clean Air Act standards was "one of the most fractious environmental debates" of the 1970s.[8]

The "fractiousness" of the debate, however, also opened new opportunities.

According to Fox, the new responsibilities—for which no mechanism existed for implementation—

served as an incentive to both parties [the government and industry] to find a mechanism for addressing some long-standing problems:

Inefficiencies: Unnecessary duplication of substantial research costs and efforts by both public and private sectors.

Inequities: Research requirements that place a burden on small manufacturers.

Lack of consistency and comparability: Use of inconsistent research methods that make cross-checking difficult.

Lack of credibility: Public suspicion that the contending parties skew data to serve their own interests.

Poor use of research facilities and personnel plus administrative delays.[9]

THE BROADER POLITICAL AND ECONOMIC ENVIRONMENT FOR CHANGE

Fox is clearly right in identifying these specific issues. However, an examination of the political and economic environment during the late 1970s surrounding the automotive arena also provides other convincing arguments about why a new institute was considered in this regulatory area and not in others where many of the problems are ostensibly similar. Three factors, in addition to the legislative impetus, seem important:

- The state of the American automobile industry in 1978–1979
- The structure of the automotive industry and its ability to change
- The nature of the general debate over the ''new social regulation''

It is critical to remember the reemergence of gas lines in 1979 and the impact of the fuel crisis on the domestic automotive industry. American car makers were increasingly faced with foreign competition. The imposition of significantly tougher emission standards, while applied across the board to foreign and domestic vehicles alike, seemed to American car and truck makers to be simply one more burden to carry. Not only would tougher emission standards raise the price of American vehicles, but the fear was that it would do so at the expense of the American work force through the loss of market share. Consequently, the economic stakes of the debate around emission standards were high.

In the diesel engine industry, the new responsibilities seemed to be not simply burdensome, but an issue of survival. If diesels were found to be per se a significant human health risk, several companies, it was feared, would simply close. In this kind of critical situation, it is not surprising that engine manufacturers were reluctant to have their fate wholly in the hands of a governmentally sponsored research apparatus whose quality and motives were not always trusted.

Even in this environment of heavy economic stakes, however, the structure of the automotive and engine industries provided the flexibility necessary to imagine "switching" rather than "fighting." Unlike some other American industries (one thinks of the electric utility manufacturers), the automotive industry is constantly changing its basic product to meet consumer demand and to stimulate additional demand. With sufficient advance warning about new findings and new requirements, the industry can change without endangering its basic markets. This "retooling capability" makes the industry relatively more open to new knowledge than others. This is not to say that industry scientists approach health effects research with a desire to find new problems. However, the nature of the industry does provide senior managers with an incentive to be at least neutral to new knowledge, provided that work is seen as credible and unbiased. This openness at senior levels was critical to the establishment of the Health Effects Institute.

Beyond the economics and structure of the automotive industry was the changing mood of the nation toward regulation. The late 1970s were watershed years in the debate over the new social regulation. In 1977 the Congress for the first time overturned a significant decision of the health regulatory agency by permitting saccharin to remain on the market. The success of interest groups in persuading the Congress to override the Food and Drug Administration's decision emboldened critics of traditional toxicological approaches and set into motion the debate on the merits of animal testing and risk assessment that persists. Academic approaches to regulatory analysis, most notably embodied in the concepts of benefit-cost analysis, began to emerge from the campus onto the national political scene. President Carter, following up on initiatives begun during the Nixon and Ford administrations, issued an executive order (E.O. 12044) requiring substantial economic analysis prior to new regulations. The environment for considering new changes and alternatives to traditional regulation transcended the automotive industry to all facets of health and environmental policy.

In sum, some very specific legislative responsibilities in the Clean Air Act were coupled with the structure and economics of the automotive industry to produce an environment within which change could be considered. These specific factors interacted with larger societal concerns to generate the necessary momentum to overcome obstacles to institutional development.

THE SCIENTIFIC CONTEXT FOR CHANGE

In addition to the legislative, economic, and political factors that led to HEI's development, there were significant scientific problems that made (and make) the call for "sound science" a reasonable one. Despite the politicization of animal experimentation caused by the saccharin debate, there were and are substantial problems in extrapolating from animal to human experience. For example, mechanisms of action for toxic substances differ from rodents to humans in some cases, and those differences need to be understood. Problems of mobile-source health effects often involve complex chemical mixtures, varying human exposure levels, and populations with different biological sensitivities.

At the level of scientific organization, it was also apparent that much existing "regulatory" research often focused exclusively on the research necessary to determine specific standards rather than on identifying the most important research needed to understand broader health effects issues. Conversely, much basic research ignored the important regulatory questions that were facing policymakers at any given time. There were, then, significant scientific questions outstanding that provided the basic substance around which an institution could be formed.

BUILDING A BRIDGE: THE FOUNDING OF THE INSTITUTION

By November 1984 Administrator William Ruckelshaus could cite the Health Effects Institute as a prime example of "the emergence of 'bridging institutions' between industry and government and even between industry and the green machine" (environmentalists).[10] Ruckelshaus went on to say that such institutions provided an environment in which "the old hostilities are breaking down, and old grievances are giving way to a spirit of cooperation, a new willingness to listen."[11]

This optimism about this new "bridging institution" was born of substantial groundwork. At times, early on, it appeared that the bridge would collapse, as the transition between Democratic and Republican administrations caused predictable problems of distrust and a resorting of priorities. This is not the place for an extended chronology.[12] It is important, however, to identify several characteristics of the HEI foundation process that probably would assist in the development of any new institution that proposes to underpin the federal health and safety regulatory structure.[13]

Top-down Development

From the beginning, the concept of HEI had support at senior executive levels. Henry Schacht, president of Cummins Engine, Thomas Murphy, chairman of General Motors, and Douglas Costle, administrator of EPA, were all personally involved in the developmental effort. The involvement at this level was important in overcoming natural and substantial bureaucratic inertia within both the government and the industry. The depth of commitment from this group is perhaps best epitomized by Schacht's statement at the press conference during which HEI was announced: "The Health Effects Institute is a quest for truth to the benefit of all of us, the general public, who live and breathe the environment with which this planet is blessed."[14]

The Presence of Full-Time Public and Private Entrepreneurs

In the private sector, a vice president for the Cummins Engine Company worked nearly full-time for a year to develop the concept, enlist the appropriate support, secure a Board of Directors, and in general guide the effort. In the federal government, a seasoned career manager within the Environmental Protection Agency played essentially the same role. This entrepreneurial behavior, motivated by interest rather than fiat, was critical in overcoming problems.

The Early Identification of Senior People to Lead the Organization

It is clear that the early willingness of former Watergate Special Prosecutor Archibald Cox to agree to chair any new institution was a coup of substantial magnitude for the still-fledgling concept. The reputation of Cox for independence and integrity was not only in tune with the goals of the institute, but Cox himself forced people to take the entire enterprise more seriously. For some time, Cox and his fellow directors, Donald Kennedy, president of Stanford University, and William Baker, then president of Bell Laboratories, were the Health Effects Institute.

The Presence of a Well-Developed Written Plan and Proposed Bylaws

In addition to the "people" components previously identified, it was undoubtedly important that everyone involved had a written plan from which to work. This plan and subsequent bylaws attempted to deal with a host of administrative and management problems that had been anticipated by the entrepreneurs. Quite clearly, the necessity to have a plan argues strongly for a condition of success laid out by Fox in 1981: "The proceedings should allow sufficient time to create a mood for exploration and compromise and a commitment to a set of operating rules."[15]

From the perspective of hindsight, changing environmental conditions provided the opportunity for people at a variety of levels in the public and private sectors to bring substantial energy to bear. The confluence of the right people in the right environmental situation is a prerequisite for success in any future effort that might emerge from HEI or independently in other regulatory situations.

A STRUCTURE TO MEET THE PROBLEMS

It is useful to return to Fox's formulation of the basic problems facing both EPA and the automotive industry in order to see how the Health Effects Institute was ultimately structured to meet these issues.

Fighting Inefficiency and Unnecessary Duplication

Both the government and industry submit their research priorities to the institute on a yearly basis. The institute's staff weigh these competing priorities and recommend a program structure to the institute's Health Research Committee. The committee then independently decides upon a single program and solicits proposals from the entire research community. The institute makes no distinction between the funds provided by the government and those provided by the automotive industry (except in the case of foreign funding, where government funds are not permitted). The process tries to ensure that the highest-priority areas are chosen for research, although the institute often takes a somewhat different slant on research problems than originally suggested by the sponsors. In a recent meeting in Asilomar, California, for example, *Science* magazine reported the following reactions to requests for short-term changes in the HEI research program: "HEI's leaders say they will not be deflected from their deliberate plans for carrying out research which they think is important, although they do poll sponsors to adjust their priorities."[16]

Addressing Inequities in Research

The funding structure of HEI, as well as its commitment to openness, ensures that all research resides in the public realm. The presence of public funds in the structure ensures that all research will be available to sponsors and to the public. The motor vehicle industry itself has developed a formula for paying its share of operational and research costs that is tied to market share on a quarterly basis. At root, however, is the commitment to share all information fully with all sponsors. The smallest corporation receives the same information as General Motors.

Developing Consistency and Comparability

Issues of consistency and comparability are not easily dealt with. In all of its materials, HEI emphasizes what it calls "quality science from definition to review."[17] In much of its approach, the institute has attempted to define a process of "quality assurance."[18] This quality assurance process does not attempt to force research proposals into a common theoretical or implementation framework. Rather, it attempts to measure each proposal and its implementation against the highest standards currently in use in the scientific community. In addition, it attempts to use a consistent process for obtaining research, modeled primarily upon the grant review process of the National Institutes of Health.

Only in a few studies has the institute attempted to force a structure upon research. These are primarily in large, multicenter studies whose outcomes are likely to have direct relevance for acceptable regulatory standards. The institute's largest undertaking to date has been in the area of carbon monoxide research, where the scientific data concerning health effects of low levels on individuals with cardiac problems has been in question.[19] In this instance HEI not only solicited proposals but has directed the study from its headquarters.

While it seems clear that sponsor expectations included the goals of comparability and consistency, it may be that these expectations are mismatched with the realities of American science. In speaking with the members of the HEI scientific committees, it is clear that there is no single accepted way of doing science in this area, and that one has to be careful in imposing a single set of standards.

The Problem of Credibility

In examining the record of HEI's development and implementation, it becomes apparent that the issue of credibility has been given top priority. The founders of HEI seem to have believed that credibility resides in the interaction of individuals, sources of funds, and processes for using those funds. Each deserves some discussion.

Individuals

The selection of both the Board of Directors and the HEI scientific committees was based on the twin criteria of scientific excellence and a reputation for integrity, with the latter predominating at the board level and the former in the scientific committees. The selection of Archibald Cox as chairman was clearly symbolic of the desire to convey integrity and independence. Similarly, the selections of William O. Baker, a pioneer in American industrial science, and Donald Kennedy, a former Food and Drug Administration commissioner, were designed to allay the natural fears of constituencies that interests would not be considered, without turning the organization into another representative political one. The selections of Walter Rosenblith from MIT and Robert Levy from Columbia as heads of the scientific committees conveyed a commitment to scientific excellence. Importantly, there was not heavy emphasis on knowledge of environmental health matters in the selection process. Since the original formation of the Health Research Committee, some additional emphasis seems to have been placed upon expertise in the toxicology and clinical aspects of emissions research, although the general orientation still seems to be toward a disciplinary approach to expertise.

In reviewing the reasons for making initial scientific choices in favor of disciplinary excellence rather than environmental health experience, it is clear that the problem of credibility played an important, if not dominant, role. The founders of the institute did not want individuals involved who were known to have either "environmentalist" or "industry" axes to grind. Rather, there was a strong desire to develop a new community of scientists who had the ability to make contributions to the area, but who were able to approach problems without any previous ideological or issue commitment. Whether the benefits of this choice outweigh the obvious costs in terms of specific knowledge is still an issue.

Source of Funds

It is not clear whether the current fifty-fifty split in HEI funding between industry and government was originally intended. The former deputy assistant administrator of EPA who was originally involved in the establishment of HEI says that it was not; others in EPA dispute this.[20] Whatever the original intention, it seems clear that the joint and equal nature of funding has contributed to the independence of the institute. On a day-to-day basis, the joint nature of funding can make life difficult for HEI staff as they attempt to assuage often-competing demands. In the long term, however, it is precisely the conflicts between demands that permit the HEI Health Research and Review Committees to make independent choices. In this independence rest the fundamental criteria in American society for credibility. Within HEI there is no incentive for scientists to do anything but speak their minds honestly, inasmuch as any choice is likely to bring criticism from one group or another.

Whether an absolute fifty-fifty funding requirement is a necessity to obtain this independence is questionable, although the difficulties encountered by another organization, Clean Sites, Inc. (CSI), tend to argue that some substantial governmental participation is a prerequisite to independence and, consequently, to credibility. Clean Sites is founded on many of the same principles as HEI, but without the government funding. It is designed to help private companies clean up hazardous waste sites. It is still perceived by some as a creature of the chemical industry, and that perception is reinforced by its funding structure. It may ultimately be successful, but only if the financial incentives for industrial-governmental cooperation are strong.

It is absolutely clear that independence can only exist in the presence of tension between funding partners. Ironically, then, organizations like HEI and CSI may work best when both the government and industry pursue their traditional organizational mandates most vigorously. In the government's case this probably means a commitment to law enforcement but without an inherently adversarial tone.

The Research Process

Strong and respected individuals are a prerequisite for credibility; joint and equal funding clearly makes independence and integrity easier to implement and easier for the public to see. On a day-to-day basis, however, it is the quality of the research process that will determine whether HEI is a credible institution. As Douglas Costle puts it, "HEI will ultimately stand or fall of its own weight, i.e., is it viewed favorably within the scientific community?"[21]

By and large, the founders of HEI accepted the general research process undertaken by one of the most highly regarded parts of the American scientific apparatus, the National Institutes of Health. Almost all of HEI's research work is done by individual investigators, primarily at academic institutions, who are awarded contracts after a highly competitive process. All proposals are reviewed and ranked by experts in specific areas of interest and then reviewed for both quality and relevance by the Health Research Committee. The Board of Directors formally makes all approvals upon recommendations from its Research Committee. This last step enables the board, with the help of the institute staff, to review the process to ensure that no conflict of interest or issues of favoritism have crept in.

While research is under way, staff of the institute attempt to ensure that the original scope of work is adhered to, or that changes in the scope are reasonable. A regular series of site visits is maintained for each project to enable HEI staff and committee members to understand progress of work and to recommend changes. Each project is evaluated for renewal on an annual basis. Projects are audited on a for-cause and random basis by the independent auditors of the institute. At the end of fiscal year 1985, 15 percent of all projects had been financially audited. No projects had been terminated for either financial or scientific reasons.

Perhaps the most unique and innovative part of the HEI research process is its conduct of postproject peer review. From the beginning of the concept, a primary driving force for industrial participation has been the desire to ensure that science was sufficiently evaluated prior to its use in the regulatory process. This element was of particular concern to car and engine manufacturers, inasmuch as there was wide recognition that initial air quality standards, despite being very quantitative and specific, were based upon shaky scientific foundations.

To meet this scientific concern, as well as to meet the demands for integrity that were placed upon the process by Professor Cox and others, the organization of HEI was founded with a clear split between the Research Committee, which developed the research program, and the Review Committee, which conducted and would publish its initial peer review of completed projects. The notion of publishing criticism at the same time that research reports were released pushes up against the edges of acceptability in American science. The prevailing ethic is one of permitting scientists to speak their conclusions in writing, after vigorous internal peer review, but not to condition the acceptance of ideas through the joint publication of peer criticism. Whether HEI can successfully implement this new mode of scientific dialogue in the long run remains to be seen.

Improving Facilities Use and Reducing Administrative Delay

While improving facilities use and reducing delay may have been perceived as major problems in 1981, there is little evidence that the institute has addressed these problems explicitly. Implicitly, the reliance upon competitively bid projects should maximize the use of available facilities, inasmuch as investigators must be able to plan the project budget and project time prior to project award. In some few cases the institute has reached out and chosen specific institutions that seem to offer the best combination of facilities and administrative excellence to do work.

With respect to delay, the problems of cutting a new swatch in the field of environmental health science meant that HEI took a significant period of time to implement. In the first three fiscal years of its existence, fiscal years 1981–1983, the institute spent a total of only $1 million. Much effort was expended in working with the federal government to develop a grants and contracts mechanism that permitted HEI to work with a variety of institutions. Ironically, then, efforts to improve the quality of facilities use worked against the very quick implementation of the entire HEI program. One of the major questions about HEI discussed later is whether the short-term perspective of budgeteers can tolerate the time necessary to implement the range of work necessary to adequately evaluate the institution.

IMPLEMENTING CHANGE AND DETERMINING SUCCESS

The previous section briefly examined how the founders of HEI conceptually attempted to meet the major problems identified by the sponsors. It seems clear that the most substantial efforts were made to ensure that the work produced by HEI would be perceived as credible both politically and scientifically. Other problems were dealt with, but given second priority. In this section several criteria for success are developed. Although judgments concerning the value of research are inherently difficult to make, and the HEI program has only been under way for a relatively short period, it seems reasonable to ask the following questions:

• Is HEI, in fact, independent, inasmuch as independence leads directly to questions about credibility?

• Even if HEI is independent, is it perceived that way by the interested scientific and policy communities who will judge it?

• How much of HEI's work is directly "relevant" to the regulatory problems that face the automotive industry and the Environmental Protection Agency? What is the right definition of "relevance" in the environmental health context?

• Is HEI likely to produce work in a timely fashion? Will it meet the needs of the regulators? Will the regulators alter their schedules to accommodate the schedule of HEI—that is, is HEI's work seen as critical to the policy outcome? Does HEI have an integral role in the risk-assessment process? All of these factors relate to the timeliness argument.

In addition to these very specific questions, some larger issues remain: Is it possible to separate science from the adversary process, and is it desirable? If it is possible, is the model chosen by HEI the right one? Finally, can HEI influence the adversarial nature of the debate and perform the "bridging" function so fondly hoped for by its founders and advocates? These questions are addressed here.

IS HEI INDEPENDENT?

Questions concerning independence must be dealt with on two levels:

- How are decisions made? Is any coercion brought to bear in the development and implementation of the program, and if so, do the principals in HEI act on the basis of that coercion?
- Is HEI intellectually independent? Do the principals have their own ideas about the structure of research in this area, or are they creatures of other people's agendas?

Coercion and the Decision-making Process

While the line between suggestion and coercion is always a thin one, the evidence around HEI suggests that it has become more, rather than less, independent over time as the sponsors have accepted a suggestive, rather than coercive, role. At the outset of discussions, the founders of HEI made it quite clear that independence was a prerequisite to participation. Indeed, Professor Cox, according to all sources, made his participation dependent upon a promise of independence, as well as upon the joint nature of the funding.

At early meetings between the chairman of the newly formed Research Committee and the sponsors, however, it became evident that many sponsor representatives had a difficult time accepting the notion that the companies or the government did not control the institute's agenda. Even as late as the winter of 1985, it was clear that tension still existed over the amount of influence the sponsors would or should have in setting the scientific agenda of the institute.[22] However, the character of the debate has markedly changed, as both governmental and industrial sponsors have grown to understand that the Board of Directors was serious in its commitment to independence.

Ironically, the primary test of HEI's independence has come not from industry, as originally feared, but from the government. Former EPA Administrator Anne Gorsuch, or at least her chief of staff, John Daniels, seems to have been highly uncomfortable with HEI's independent role at the outset of the Reagan administration. Whether this was motivated by partisan politics or by an excessive degree of administrative zeal, the spring of 1982 proved to be a difficult one, as Gorsuch tested the mettle of the institute by trying to rein in its activities. At other levels various parts of the EPA seem to have had a more difficult time than the private sector accepting the fact that HEI is not a "contractor," subject to the whims of agency personnel. This problem has surfaced particularly in the emphasis (or lack thereof) HEI has placed on the subject of methanol fuel research. HEI's scientists simply do not see that methanol research is of great interest, given other problems and the low likelihood of rapid implementation in the United States. This has put the institute occasionally at odds with those in EPA who would like to see the use of this fuel encouraged and would like to develop the fullest possible data base. Over time, this problem seems to have been accommodated, but not without some substantial bureaucratic difficulty.

It is clear that the board and the scientific committees of HEI cannot be grossly intimidated. It is also clear, however, that the role of the staff, and particularly the executive director, is one that must accommodate attempts to influence the institute and turn those attempts to positive ends, if at all possible. Because the executive director of the institute is between two strongly contending forces at all times, it is inevitable that threats will be made and that "suggestions" for action will carry coercive aspects. The staff cannot simply accept all such attempts; alternatively, however, it is critical to know when to take such threats seriously and when to chalk them up to the cost of doing business. In an institution like HEI, therefore, it is the role of the staff to transform criticism into constructive action without unnecessarily calling into play the very independent structure that forms the board and committees of the institute. Indeed, it is this aspect of sponsor relations that makes it desirable, and perhaps necessary, that the executive director of HEI be not simply a scientist, but also experienced at the darker bureaucratic and political arts.

Intellectual Independence and HEI

Several years into the HEI research program, it is still difficult to say how intellectually independent the program is. The Research Committee members and the board are, as Professor Cox pointed out to Ron Fox in 1981, "fiercely independent."[23] Indeed, the reaction to the institute's first set of research proposals by the sponsors was that HEI was going too much its own way and ignoring sponsor requests.

Possibly in response to this initial criticism, and possibly because of the energy required to remain intellectually independent, the HEI program in the mid–1980s began to resemble in many respects the collective priorities of the sponsors. To be sure, the requests of the automotive sponsors are often different from those posed by EPA.[24] (A prime example of this is the continuing emphasis placed by EPA on methanol and aldehydes as opposed to the heavy emphasis placed by industrial sponsors on other issues.) There can be no doubt that the HEI Research Committee takes solace in these differences, for it is precisely in the difference between industrial and governmental agendas that the institute can find its own way. In spite of these differences, however, an observer could argue that HEI has been too reactive. The institute's principals seem to recognize this problem and have taken steps via a significant long-term planning effort to stimulate the intellectual climate.

IS HEI PERCEIVED TO BE INDEPENDENT AND CREDIBLE?

Epistemological differences between so-called "perception" and "reality" are much discussed in political circles these days. Indeed, the question of what "spin" is on an issue often seems to dominate policy. The founders of HEI unquestionably believed the following:

• There is a difference between facts and values, and, as Henry Schacht of Cummins Engine has been heard to say, "not every fact can be negotiated."

• Decisions about the conduct of science are best made by scientists, and one needs to know something about the disciplines involved in the health effects of auto emissions in order to make the most useful contribution.

• The combination of people of integrity and a reputation for independence with scientists of the highest quality will produce a research program that is policy neutral and oriented toward the production of new knowledge.

• The imposition of an internal peer review process—together with the other mechanisms described earlier—would nearly guarantee a program whose results people would find compelling. As former Administrator Costle has said, the aim was to develop a set of "friendly facts."[25]

As the 1980s closed it is now possible to say that the institute appears credible. However, it is possible to point out several problems that will have to be faced if the institute is to maximize its credibility.

Relationships with the Public-Interest Community and the Press

Despite the evident turn of Americans toward a more conservative view of the world, it is clear that the media are very influential in forming public impressions. In turn, the media in environmental matters often turn to the public-interest community, on the grounds that these individuals have no financial interests to grind and that this purity breeds objectivity when the public health is involved. Because the HEI model, as expressed here, relies heavily upon demonstrated "expertise," it is possible that conflict will occur between the prevailing dogma of the media and public-interest community and the values propounded by the institute.

Specifically, the dogma and the value system are likely to clash over the issue of funding sources. Despite the fifty-fifty nature of HEI's funding, some in the public-interest community will disparage HEI either on the grounds that it takes money from the industry (thus inevitably prejudicing its work), or that it is diverting funds that the government itself would use for work in similar efforts. The institute will be especially vulnerable to these arguments if a substantial part of its research work fails to uncover evidence concerning health hazards.

Consequently, unless the institute works very hard to ensure adequate understanding of its efforts by the media and the public-interest community, the likelihood of the wrong "spin" being put on its work will persist.

Criticism from within the Scientific Community

Even if the institute is able to overcome any problems it may have on the overtly political front, it may face a challenge to its credibility from within the scientific community over its review process. To reiterate, that process, which requires a substantial internal peer review and the publishing of that review along with the scientific paper, is radical in American science. It has been devised by the sponsors and the Board of Directors as a mechanism for buffering the use of the work in the regulatory arena and for ensuring that nonscientists can adequately interpret results. It runs the risk, however, of alienating some segments of the scientific community. That alienation could invite subversion from within.

To counteract this possibility, HEI is attempting to couch the language of its reviews very carefully and to work closely with the investigators it funds to help them report their work in the most complete possible form. The aim, obviously, is to produce work that makes HEI investigators happy while meeting quality requirements that are immediately useful for risk assessment. Many of these fears and problems can be dispelled only through the production of a great deal of research, the reporting of much of which is still in the future.

THE QUALITY OF HEI RESEARCH

Almost everyone interviewed in connection with this chapter admitted that the litmus test of HEI will be in the quality of its science. That, ironically, is the aspect of the institute that is least under the control of the founders, the current principals, and even the HEI staff. By committing itself early on to the competitive funding of investigator-initiated grants, HEI took a leap of faith in American universities and in the mode of funding that had brought greatness to the National Science Foundation and the National Institutes of Health (NIH). Any attempt to evaluate the quality of HEI-sponsored science at this point would be premature, but some analysis indicates that the quality is comparable to that undertaken in other parts of the biomedical sciences, but perhaps with a higher degree of variability. The analysis is based on the following points:

- HEI awards are highly competitive, although a higher percentage of awards are made than by the NIH. In its first three years of funding, HEI made contract awards in approximately 15 percent of all applications.[26] This is competitive with, but higher than, the overall rate of approval by NIH study sections.[27]
- The so-called "pay-line" for HEI contracts, that is, that "priority" below which money

is provided, is declining over time and is now at the level approached by the NIH in 1980. Since that time the NIH has become extremely stringent as the administration attempts to reduce the total number of grants. The upshot of this is that the average HEI grant might have trouble being funded by NIH today, but this is not necessarily a reflection on the overall quality of investigators in HEI.[28] Indeed, one can argue that the presence of institutions like HEI is more important than ever today in the face of constricting federal resources.

• As an unknown institution when it began, HEI funded a number of studies initially that would undoubtedly not be funded today and will not be renewed. As HEI progresses, the quality of investigators attracted to it is improving, and this is borne out in declining mean and median average priority scores for funding (lower scores are better in this parlance).

• By HEI's own admission, the status of the science in this area has not been high. The number of investigators with primary interest in the direct health effects of auto emissions is small, and the overall quality of the science is suspect. If that had not been the case, the motivation for HEI would, in some fashion, have been nonexistent. After several years of funding, the institute has attracted some investigators who have never worked on problems related to air quality before but who are simply good biomedical research-ers. The process of forming a new community of scientists will undoubtedly result in some very good studies, along with some that do not work out.

On balance, there is no reason to predict that the quality of HEI science will be worse than the area had before. To the contrary, there are a number of reasons—ranging from the quality of scientists involved in making the decisions to the mode of decision making—to expect that over time the science will improve dramatically.

One final note is warranted on the issue of quality. To many non-scientists, the word "quality" is synonymous with "products" or "results." Too often, people have the conception that a single well-conducted study can solve a prob-lem. Every Thursday, the publication of results in the *New England Journal of Medicine* encourages the notion of the progress of science. It is much more often the case that "quality" science involves the elaboration of an old problem and provides new ways of looking at it; indeed, quality science may even be the first fumbling attempts to look at a new problem. Consequently, one must be wary of any glib definition of "good science," particularly one based upon a stylized view of what science is about.

HOW RELEVANT IS HEI?

The question of relevance is not simple. Answers involve a discussion of distinctly different perspectives between scientists and regulators, as well as differences with the scientific community about the best way to generate new knowledge.

There is no question about the mission of HEI. It is clearly meant to be, as former Administrator Costle put it in the press conference announcing it, "the major source of studies on the health effects of motor vehicle emissions in the country and a key to improving EPA's capacity to tie its regulatory action to sound health effects findings.''[29]

In a speech to the Society of Automotive Engineers in 1984, the first executive director of the institute, Charles Powers, discussed the basic conflict in evaluating the term "relevance":

When the mission is improved regulatory decisions, the objectives of a mission oriented research program, like HEI's, are always on the verge of being subverted by the Sherlock Holmes approach of the scientist for whom pathogenesis is more salient than filling or creating chinks in a criteria document.

But there is, equally, a very real danger in thinking that environmental science can focus primarily on the regulator's most recently announced worry or baseline for a standard. There is a danger when the regulator's question becomes the lead strand. The inevitable result of such a focus is to begin a pattern of repeated studies hovering in concentric circles around the regulatory decision-point. What should, scientifically, be only a clue becomes a battleground where all contenders pitch their tents. The measurement begins— like how long before the next criteria document and what study can we do to be finished before then. The search for methods begins to focus not on what might best pursue the clue but on how best to standardize on the method which created the clue.[30]

If it is true that the regulators feel compelled, sometimes by law, to focus on a particular study that produces results, it may also be true that scientifically that study is irrelevant to understanding the problem, either because it presumes that other work has been done that is unavailable, or because the problem being pursued is, in some fundamental way, trivial. In attempting to operate within this inevitable tension between science and near-term demands, the HEI program has tended to become more oriented toward ensuring that some of its projects can be clearly seen by all to affect regulatory decision making.

The initial HEI research program, which was first advertised in the spring of 1982, was very oriented toward the exploration of fundamental scientific problems. Since then, particularly with its emphasis on carbon monoxide, the health effects of gasoline vapors, and a generally increasing emphasis on those aspects of mobile-source air pollution not regulated by national ambient air quality standards, the HEI research program has tended, on the margin, to take on a more applied cast. It still remains to be seen, however, whether this emphasis will do more harm than good, as it tends to make the issue of timeliness discussed later more important. With a relatively basic program, it is easy to deflect issues of timing. Once it is clear, however, that the question of the test of timeliness must be applied to a research program, the test of relevance becomes more important.

Even within the scientific community, however, the issue of relevance is not simple. There are some within the HEI scientific tent who believe that all of HEI's studies must be related very directly to the risk-assessment model that has been put forward by the National Academy of Sciences.[31] In essence, this view would require that all investigators in an institution like HEI be able to identify, prior to beginning work, how their science is likely to contribute to the evaluation of human risk. Other scientists believe that this amount of formalism is intellectually dishonest or, at least, impossible. Science, they argue, is constantly serendipitous; it is often impossible to know whether or where science is going to make a contribution. Because this is true, it is important not to emphasize relevance too greatly in making decisions about research funding. On the margin, this latter view probably continues to dominate in HEI, although there is a sense of increased flexibility as the HEI's Research Committee evolves.

EPA's clear expectation is that the program will continue to balance short-term research needs against a broader program of methodological development. In the near term it is universally recognized that the outcome of HEI's multicenter carbon monoxide study is critical to how the institute is viewed. That study is expected to focus substantial attention on an issue of some consequence, inasmuch as it directly underpins the national ambient air quality standard for carbon monoxide. How the results of this study are viewed and used will greatly determine the future of the Health Effects Institute.

WILL HEI BE "TIMELY"?

When the development of HEI is examined, the length of time it has taken to develop the research program is striking. In the first three fiscal years of the organization's operation, only $1 million was spent, mostly in setting up the organization. Research was not under way until the late spring and early summer of 1983, and it was not until fiscal year 1985 that the institute could be said to be operating at nearly full capacity. In short, nearly five years elapsed from the time the institute was formally announced until the first research was ready for reporting.

During that time the political environment around the Health Effects Institute changed dramatically. Not only did the politics of regulation undergo a revolution, but so did the relationship between the government and the automotive industry. The importance of diesel engines, which one can argue was the primary driving force behind the development of the institute, declined in light-duty vehicles in the United States, although substantial interest remains in Japan. The focus of environmental policy shifted from mobile-source air pollution to hazardous waste sites, to toxic air pollutants from stationary sources.

At one level the results of all these changes were beneficial to the institute. They permitted HEI to grow in an atmosphere where neither the staff nor the scientists had to respond to the "pollutant of the week" syndrome. The relatively quiet environment in the mobile-source area allowed the organization to come together and permitted the development of something resembling a community of scholars.

There are also costs. Because neither the public nor the regulatory community have seen the fruits of HEI's labor, there is a tendency to ignore its existence. The institute will now have to elbow its way into the policy process and develop de facto influence to accompany the de jure responsibilities that have been delegated and expressed by a variety of people. Its leadership will have to ensure that the agency and the industry take its work seriously and are even willing to have their timetables altered by the time it takes to put HEI's work on the table.

This necessity to be timely will require HEI's leadership to lobby. However, the raison d'être of the institute is to eschew formal lobbying, inasmuch as direct involvement in the political process can be seen to run counter to the institute's mission. How the organization deals with this ambiguity and how successful it is in making its work fit into the decision-making timetables or in altering the timetables will determine to a great extent the external perception of whether the institute is a success. On a very practical level, it is important that HEI be perceived by budgetmakers as timely when they attempt to allocate scarce resources.

CONCLUSIONS

When HEI is evaluated in terms of a number of specific criteria, it is probably fair to conclude that the institute is promising, but that the results are not all in. The institute seems well situated to be credible and independent, but the conflict between the requirements of the "expertise" model it employs and the political environment may create problems despite the presence of individuals such as Archibald Cox. The quality of the research, as a body, is surely not worse than that of any other comparable institution; to the contrary, it is improving as it goes along and in many respects is derived from a superior and well-proven process. It offers real possibilities for the environmental sciences. The issue of relevance is more complicated, and the differences in weltanschauung between the scientific and policy communities probably ensure that this issue will be a source of continuing tension. Finally, and perhaps most importantly in the near term, there is the question of timeliness. Will HEI produce sufficiently and be judged as useful before the patience of the sponsors and the budgeteers expires? The answer to that lies in the skill of the entrepreneurs who founded the institution and in the willingness of many people, from Archibald Cox to William Ruckelshaus, to stand up for an idea that they believe is right.

If it is true that the success of HEI depends largely on the patience of its sponsors, and if that patience depends upon the willingness of its backers to stand up for the concepts, then the ultimate power of the HEI lies in the answers to several questions:

- Is it desirable to separate science from the adversarial process?
- Is it possible to separate risk assessment and risk management?

• Is it desirable for public and private institutions to work together on problems in the environmental arena?

FINDING "TRUTH": SEPARATING SCIENCE FROM ADVOCACY

This chapter has spent some time arguing around the question: What is science? If we believe that "science" is some combination of "fact," that is, what we know with a high degree of certainty, and "inference," that is, that which we suspect with some knowledge but where the range of uncertainty is higher, then it seems clearly desirable to separate science from advocacy. This is because the interesting places in the science are the places of inference. If we are advocates of a position, we will be more or less likely to accept uncertainty, depending upon our viewpoint. If, however, we are neutral, that is, we can see the level and kind of uncertainties with some objectivity, this objectivity clarifies the scientific choices that remain to reduce uncertainty, and enables one to ask society the critical questions about whether it wants to make the investments necessary to reduce the field for inference about human risk. To sum up the issue simply, the removal of science from the adversarial process enables the ultimate adversarial process to take place on a more informed and honest level. In a democratic society that clarification can only be a good thing.

Is the Separation Possible? Risk Assessment and Risk Management

Given that it is desirable to separate science from policy conflict, is it really possible? It is clear that the prevailing point of view within the government today is that it is. The separation between so-called "risk assessment" and "risk management" is precisely an attempt to isolate facts from values and to introduce some order and consistency into what has been a disorderly process. Despite the "ideology," however, one can still fairly ask whether it is really possible to separate the two processes. At some level the amount of resources channeled into any case of risk assessment is already determined by the ends to which one wants to use the evidence. Specifically, it may be much more important to do extensive exposure analysis on a particular chemical if the expectation is that one is going to substantially limit its uses based upon the quality of that information. Clearly, then, there can be an overlap between the process of managing and assessing risk.

Simply because there is an overlap between facts and values in some cases is no excuse for arguing that there is no difference. Unfortunately, this is a logical fallacy that easily creeps into any discussion of science-based regulation. It is here that the notion that "not all facts can be equally negotiated" becomes relevant, and it is here that the Health Effects Institute takes on life. Only by pounding away at the difference between facts, inferences, and values can HEI attempt, admittedly on the margin, to change the character of cynicism that so pervades much of the public policy process. When any member of the policy community attempts to torque the scientific information to meet policy goals, it is the mission of the HEI to report this and, if possible, provide some discipline in the regulatory process.

At base, then, the answer to questions about the possibility of separation is "more often than we have." It is a useful myth to report an unqualified "yes," because so many cynics are willing to say "no" in order to meet their own policy ends.

Blurring the Sectoral Lines

HEI is unique in how it is funded. This model of joint public- and private-sector funding is, on its face, an ideal use of resources. Theoretically, at least, it permits the allocation of resources to the most difficult problems without duplication.

The issue remains whether such joint ventures are inherently collusive. Despite all of the process protections, can we ensure that the "public interest" is not left out of the ultimate products and decisions? At one level, the answer is surely no. It is only by the careful selection of leadership that one creates the kind of atmosphere in which collusion is not an acceptable part of the organization's culture. In HEI's case the presence of someone like Archibald Cox, with his history, clearly tipped the political balance in favor of the institute. The original idea encountered substantial resistance on Capitol Hill in the person of then Senator Edmund Muskie precisely because of concerns over the inherent nature of public-private partnerships. The fact is, of course, that there are not many people in our society with the cachet and portfolio of a Cox. To the extent that elaborate public-private cooperation requires the presence of such distinguished people, joint ventures are inherently likely to be in short supply.

Despite this pessimistic view, there are reasons to believe that these kinds of joint ventures are precisely what the American economy and polity need. Our culture simply does not value public service to the extent that many of the best people are willing to work on a continuing basis with the government. If we are to harness the expertise necessary to deal effectively with problems of the environment or, indeed, other serious societal problems, we simply must find ways to bring private-sector expertise to bear in a credible fashion. Institutions like HEI that are not merely billed as independent, but are so in fact, must form a major place where expertise can come together to solve, rather than exacerbate, problems.

As this chapter has demonstrated, the tensions within these institutions are great. The inherent distrust between public and private sectors must be managed without burying it. The managerial and personal skills of the staffs of these institutions must be exceptional, at least until the prevailing adversarial culture is controlled. The substantive scientific problems in making progress are great;

if they were not, solutions would have been arrived at already. In sum, the combination of high tension, high stakes, and tough problems that spawned an institution like HEI make the probability of ultimate success less than optimal. HEI is an attempt to change prevailing norms, and that is always difficult. In the absence of further attempts, however, we may be doomed to a gridlock in which individual players feel better protecting their fiefdoms, but in which the public interest is really not protected. Far from leaving out the public interest, then, institutions like HEI provide one of the best hopes for serving the public interest.

NOTES

1. Health Effects Institute and Jellinek, Schwartz, Connolly, and Freshman, Inc., *The Clean Air Act: A Primer.*

2. J. R. Fox, "Breaking the Regulatory Deadlock," *Harvard Business Review*, September–October 1981, pp. 97–106.

3. Ibid., p. 101.

4. Section 202(a) of the 1977 Clean Air Act Amendments.

5. Section 202(c) of the 1977 Clean Air Act Amendments.

6. Interview between Douglas Costle and Lisa Baci, April 19, 1985.

7. Interview with Michael Walsh, former Deputy Assistant Administrator of Mobile Sources, April 10, 1985.

8. Costle and Baci interview.

9. Fox, "Breaking the Regulatory Deadlock," pp. 101–2.

10. William Ruckelshaus, Speech before the World Industry Conference on Environmental Management, November 14, 1984.

11. Ibid.

12. Thomas Grumbly and Stanley Blacker, "The Health Effects Institute and the Future of Regulatory Science," U.S. Environmental Protection Agency, Contract 68–01–4038, Purchase Order 17256, May 1984.

13. Ibid.

14. Henry Schacht, Transcript of Proceedings, National Academy of Sciences, December 12, 1980.

15. Fox, "Breaking the Regulatory Deadlock," p. 104.

16. "Health Effects Institute Links Adversaries," *Science*, February 14, 1985, p. 738.

17. Health Effects Institute, "The Health Effects Institute: A New Approach to Regulatory Science," January 1985, pp. 14–20.

18. Health Effects Institute, "Applications to the U.S. Environmental Protection Agency for Federal Assistance," July 1984, p. 4.

19. Letter from Archibald Cox to the Environmental Protection Agency, September 19, 1985.

20. Interviews with Michael Walsh and Stanley Blacker, EPA official, April 10, 1985.

21. Personal interview, April 10, 1985.

22. "Health Effects Institute Links Adversaries," p. 738.

23. Fox, "Breaking the Regulatory Deadlock," p. 102.

24. See HEI internal documents for 1984.

25. Interview with Douglas Costle, April 10, 1985.

26. Information supplied by the Health Effects Institute.

27. "More Competition, More Disquiet," *Nature*, vol. 315, May 9, 1985, p. 88.

28. Ibid.

29. Douglas Costle, Transcript of Proceedings, National Academy of Sciences, December 12, 1980, p. 10.

30. Charles Powers, speech to the Society of Automotive Engineers, May 23, 1984.

31. Committee on the Institutional Means for Assessment of Risks to Public Health, *Risk Assessment in the Federal Government: Managing the Process*, Commission on Life Sciences, National Research Council, National Academy Press, 1983.

CHAPTER 5

UNLEADED GASOLINE VAPORS

Susan Egan-Keane, John D. Graham, and Eric Ruder

Since the late 1970s the U.S. Environmental Protection Agency (EPA) has identified refueling of motor vehicles as a significant source of air pollution. Gasoline vapors are displaced from the vehicle fuel tank by incoming gasoline during refueling. People who fill their own gas tanks are quite familiar with the odor of gasoline vapor.

Refueling emissions contribute to the national ozone ("smog") problem. While gasoline itself is a complex chemical mixture, refueling emissions consist almost entirely of hydrocarbons. In the presence of sunlight, hydrocarbons mix with other pollutants in the air to form ozone (and other photochemical oxidants). EPA has already set maximum permissible ambient concentrations for ozone in order to protect the public from potential adverse health effects and to minimize damage to vegetation and materials. Since much of urban America has not yet complied with EPA's ozone standards, all sources of hydrocarbon emissions—including refueling vapors—have been targeted by EPA for control.

Concern about refueling emissions became more urgent in the early 1980s when laboratory scientists discovered that wholly vaporized unleaded gasoline causes tumors in rats and mice. Risk assessors at EPA used the animal data to project cancer risk in humans exposed to refueling emissions. Since about two-thirds of all the gasoline pumped into motor vehicles in the United States is done at self-service pumps, the potential cancer risk from exposure to gasoline vapor is widespread among the public.

Two approaches to controlling refueling emissions have been considered by EPA. Equipment can be installed at gasoline stations to recover the displaced gasoline vapors and return them to underground storage tanks. These so-called Stage II controls, which are used already in California, St. Louis, Missouri, and the District of Columbia, are a natural extension of the recovery systems used at bulk terminals to capture vapors from returning tank trucks (so-called Stage I controls). The major alternative control strategy is to design a refueling vapor control system into new cars and trucks. So-called on-board control systems would capture vapors during refueling and store them in a large charcoal-filled canister that is purged to the engine for combustion during vehicle operation. Implementation of either Stage II controls or on-board systems might be accompanied by regulations on the volatility of fuels, another hydrocarbon control strategy.

The politics of regulating refueling emissions are more subtle than the typical battle between business interests and environmentalists. Here the forces in the business community are unified in opposition to EPA's position on cancer risk yet divided on the proper choice of control strategy. Petroleum interests have done little to resist EPA's apparent inclination toward on-board controls, while motor vehicle interests have advocated controls at the pump (Stage II) as more cost-effective than on-board systems. Meanwhile, the more ardent environmentalists seek adoption of Stage II controls as the short-run policy and on-board systems for permanent control.

The policy choice between "control at the pump" and "on-board control" hinges to a considerable extent on the scientific plausibility of EPA's cancer risk estimates for refueling emissions. If the cancer threat is groundless, the policymaker might be inclined to order Stage II controls in those regions of the country suffering (or predicted to suffer) from excessive ozone levels. But if the cancer threat is significant, a nationwide on-board control policy might look attractive. The stakes in the policy decision are obviously large: public health, regulatory compliance burdens, and the distribution of control costs among sectors of the economy.

This chapter examines the processes employed by EPA to obtain independent scientific advice on the potential cancer risks of refueling emissions. We focus on the workings of the agency's Science Advisory Board and the Health Effects Institute, an independent research organization funded equally by EPA and the motor vehicle industry. To a lesser extent we also describe the influence of the Chemical Industry Institute of Toxicology, an organization funded wholly by the chemical industry to perform basic toxicology research. We describe how these institutions operated in this case, what their contribution was, and how EPA and other parties reacted to their work.

HISTORY OF EPA INTEREST IN REFUELING EMISSIONS

In the early 1970s EPA adopted national ambient air quality standards (NAAQSs) for ozone to protect the public health with "an adequate margin of safety."[1] The current standard, which was set when EPA loosened the standard in 1978, requires that the one-hour concentrations of ozone in excess of 0.12 parts per million (ppm) occur on no more than one day in a calendar year.[2] The Clean Air Act of 1970 originally required all areas of the country to comply with the 0.12-ppm standard by December 31, 1982, but this deadline was extended to December 31, 1987, for states with serious nonattainment problems. Congress considered an additional extension in 1988. Meanwhile, EPA reevaluated the scientific basis of the 0.12-ppm level in light of additional research findings.

EPA determined in 1987 that seventy-three urban areas with 100 million residents are currently exceeding the 0.12-ppm standard.[3] Although dramatic progress has already been achieved in controlling hydrocarbon emissions (exhaust and evaporative) from mobile sources, EPA projects that the smog problem will actually worsen in many urban areas by the year 2010 due to economic growth and increased vehicle use.[4] To avoid this scenario, EPA officials in the 1970s began to look toward refueling controls as part of the solution.

Stage II controls are considered by many EPA and state officials to be cumbersome and difficult to implement. For instance, there are roughly 300,000 gasoline stations throughout the nation. Numerous implementation problems were reported by the several jurisdictions that tried Stage II controls in the mid–1970s. As a result, a provision of the Clean Air Act Amendments of 1977 directed EPA to consider the feasibility of on-board systems as an alternative control strategy.[5] After an extensive engineering analysis, EPA concluded in 1980 that on-board systems were indeed feasible.[6] By the end of the Carter administration, EPA appeared to be moving toward a rulemaking on refueling emissions with on-board systems as the primary method of control.[7]

The Reagan administration deferred this rulemaking plan in 1981, citing the deep recession in the auto industry and the administration's new philosophy of "regulatory relief."[8] As a result, under President Reagan's first EPA administrator, Anne Gorsuch, the refueling issue was not addressed seriously.

Regulatory interest in refueling emissions was renewed at EPA beginning in late 1982 and early 1983. Laboratory scientists supported by the American Petroleum Institute (API) found that wholly volatilized gasoline vapor caused cancer in animals.[9] This evidence was published at the same time as EPA officials came under increasing pressure from environmentalists to take action on the persistent ozone nonattainment problem. Meanwhile, the mid–1980s brought an improved economic environment for regulation and supportive leadership from new EPA Administrator William Ruckelshaus. By 1984 career officials at EPA had intensified their work on an ambitious proposal to regulate refueling emissions. This work was a collaboration of three units within EPA: the Office of Mobile Sources (OMS), the Office of Air Quality Planning and Standards (OAQPS), and the Office of Health and Environmental Assessment (OHEA). Before we examine EPA's risk-assessment process, it is useful to understand how and why gasoline vapors came to be viewed as carcinogenic.

ORIGINS OF API'S ANIMAL STUDY

As recently as the early 1970s, toxicological research was not a high priority in the petroleum industry. The American Petroleum Institute's tiny biomedical

Table 5.1
Results of API's Lifetime Carcinogen Bioassays of Unleaded Gasoline Vapors

Exposure Group (ppm)	Kidney Tumor Incidence in Male Fischer 344 Rats	Liver Tumor Incidence in Female B6C3F1 Mice
0	0/100	8/100
67	1/100	10/100
292	5/100	12/100
2,056	7/100	27/100

Source: H. N. MacFarland et al., "A Chronic Inhalation Study of Unleaded Gasoline Vapor," *Journal of the American College of Toxicology*, 3, 1984, pp. 231–48.

research program was controlled primarily by physicians and industrial hygienists; there were few trained toxicologists employed in the industry.

By the mid–1970s it was apparent that a new industrial approach toward chronic health risks would be necessary. Public concern about chemical hazards was growing, and the power of regulatory agencies such as EPA was on the rise. According to Robert Scala of Exxon, "The time had passed that we could simply generate pious statements that gasoline products are safe. We pushed for affirmative proof of safety."[10]

Scala and his colleagues on API's medicine and environmental health committee designed a two-year research program to characterize the toxicity of gasoline products. Beginning with solvents, API undertook mutagenicity studies and ninety-day animal tests. Later, when leaded and unleaded gasolines were studied, the only adverse effects discovered were attributed to lead. API planned to follow up with long-term chronic inhalation bioassays using rodents and monkeys, but the experiments with monkeys were abandoned due to insufficient funding.

In 1980 the American Petroleum Institute commissioned the International Research and Development Corporation to perform a chronic inhalation study of unleaded gasoline in two rodent species (rats and mice). According to Robert Scala, "We thought we were going into a negative study."[11] The surprising results, which were published in final form in 1984, are summarized in Table 5.1. Unleaded gasoline, in wholly vaporized form, caused an excess incidence of both kidney tumors in male rats and liver tumors in female mice. The API study was the first definitive demonstration that gasoline is carcinogenic to animals.

API staff and managers from member companies were "surprised" at these results.[12] There was a widespread fear in the industry that "EPA might force us to reconfigure gasoline."[13] Despite these fears, no attempt was made to hide or delay publication of the results. According to Robert Scala, the decision to report the results immediately to EPA "protected the industry's interests."[14] In the short run, though, API scientists were the subject of substantial criticism.[15]

EPA SEIZES UPON THE RESULTS OF THE API STUDY

The results of the API study proved to be a big opportunity for analysts at EPA who were working on refueling emissions regulation. Because the ozone rationale for on-board control systems was not persuasive to everyone, EPA analysts turned to the cancer threat as a central rationale for a regulation that was already receiving its finishing touches.

The API study was used by EPA analysts in OHEA and OAQPS as a basis for making quantitative cancer risk estimates for human exposures to gasoline vapor. The resulting staff paper, entitled *Estimation of the Public Health Risk from Exposure to Gasoline Vapor via the Gasoline Marketing System*, was released for public comment in June 1984.[16] Results from the staff paper were also incorporated into an OMS report on regulatory options that was released for public comment in July 1984.[17]

Using the criteria for evaluation of evidence developed by the World Health Organization's International Agency for Research on Cancer (IARC), the EPA staff paper concluded that unleaded gasoline is a probable human carcinogen.[18] This language, which is summarized with the IARC category 2B, has a specific meaning to cancer scientists.[19] A known human carcinogen (e.g., tobacco) is a 1. Positive animal data alone justify a rating of 2. When epidemiological information on an animal carcinogen is considered inadequate for directly assessing human cancer risk, a classification of 2B is made. An animal carcinogen with limited supporting evidence from humans is classified as 2A.

For purposes of quantifying human cancer risk, EPA relied on the standard procedures of the Carcinogen Assessment Group (CAG), a unit within OHEA. Its procedures are as follows:

- Selection of bioassay data from the most sensitive species of animals tested (in this case the data from both male rats and female mice were used)
- Fitting the bioassay data to the "linearized multistage model," which assumes that tumor response is a linear function of dose at low doses and calculates the largest possible slope at low doses that is consistent with the experimental data points
- Calculation of "dose" as cumulative lifetime exposure to the substance in question
- Making the assumption that humans and the most sensitive tested animal species will exhibit the same percentage tumor response for any given dose[20]

Based on these assumptions, EPA analysts found that each part per million of gasoline vapor (for a lifetime of exposure) causes an excess lifetime risk of cancer in humans of 2.1 per 1,000 (using mouse data) to 3.5 per 1,000 (using rat data).[21] When these results were applied to EPA's national exposure models for self-service stations, the resulting estimates of cumulative cancer incidence from gasoline vapor for the 1986–2020 period were 116 to 192 additional cases.[22] Both the staff paper and the regulatory options paper emphasized that these numbers should be considered "plausible upper limits" on the actual human cancer risk.

The EPA staff paper discussed explicitly some of the key uncertainties in the numerical risk estimates. These uncertainties included the following:

- Risk estimates are sensitive to the choice of a low-dose extrapolation model (i.e., alternative models yield different risk estimates).
- Animals and humans may not be equally sensitive to the carcinogenic effects of gasoline vapor.
- Intermittent exposure at gasoline service stations may not be toxicologically equivalent to the continuous exposures tested in the API study.
- Test animals breathed the complete mixture of gasoline vapor under laboratory conditions, whereas humans breathe only the more volatile components of the mixture, which may have different toxicological properties.[23]

Despite the uncertainties, career EPA officials were convinced that it was prudent to regulate on the basis of cancer risk. When the staff paper and regulatory options report were released for public comment, it was clear to all of the interested parties that career EPA officials were moving rapidly toward a rule-making on refueling emissions. For that to occur, however, they would need approval from the agency's Science Advisory Board and a go-ahead from EPA Administrator William Ruckelshaus.

SCIENTIFIC REVIEW BY EPA'S SCIENTIFIC ADVISORY BOARD

The Science Advisory Board (SAB) is a public advisory group of scientists that provides extramural scientific information and advice to the administrator and other officials of the U.S. Environmental Protection Agency. The board is intended to provide a balanced expert assessment of scientific matters related to problems facing the agency. Since its creation in 1976, SAB has played an increasingly active and prominent role in the regulatory process by conducting independent reviews of the quality of EPA's scientific work.

OAQPS officials submitted the staff paper on gasoline vapor to the Executive Committee of the Science Advisory Board for review. Following usual practice, the Executive Committee referred the paper to one of its standing committees, the Environmental Health Committee (EHC), which possessed substantial expertise in cancer risk assessment. The membership of this committee as of July 1984 is listed in Appendix B of this chapter.

Under the leadership of Chairman Herschel Griffin, EHC elected to hire several special consultants to supplement the committee's expertise. Jack Gray of the Upjohn Company, a pathologist, was recruited as an expert on rat nephropathy. Samuel Lestz, a mechanical engineer from Pennsylvania State University, was chosen as an expert on the chemical composition of fuels as they relate to human exposure. Ronald Wyzga of the Electric Power Research Institute was hired as an expert on the application of quantitative risk-assessment techniques. These consultants were commissioned by EHC to review the EPA staff paper, attend EHC's public meeting, and serve as ad hoc committee members on the refueling issue.

EHC's Public Meeting

Each committee member was mailed a copy of the staff paper (and some written comments from interest groups) roughly one month in advance of the public meeting. Members were asked to focus on Chapter 5.0, "Evaluation of the Carcinogenicity of Unleaded Gasoline." The other chapters, which provided background information on control strategies and policy options, were provided as context for committee members even though they were not to be reviewed by EHC. The public meeting was held at EPA in Washington, D.C., on July 25, 1984. Transcripts of the meeting were taken and are publicly available.[24]

The executive secretary of EHC, Daniel Byrd, withdrew himself from the review on the grounds that his former association with the American Petroleum Institute might be perceived as a conflict of interest. The director of SAB, Terry Yosie, assumed Byrd's role as staff to EHC.

Formal Presentations

The first part of the public meeting on July 25 was consumed by presentations from representatives of key interest groups. Major points from these presentations were as follows:

- Robert Scala of the American Petroleum Institute testified that "existing animal data are not sufficient to warrant use of quantitative risk-assessment techniques." He urged EPA to "postpone" any quantitative risk assessment until the results of API's ongoing research efforts on the rat kidney were available.

- David Doniger, an attorney with the Natural Resources Defense Council, "agreed with EPA's overall conclusion" in the staff paper, urged EPA not to wait for "all the i's to be dotted and the t's to be crossed" by researchers, and emphasized the importance of being "prudent" in protecting public health.

- Jaroslav Vostal of the Motor Vehicle Manufacturers Association (MVMA) testified that EPA's risk assessment should be revised because "the exposures to the animals in the API study are not representative of the actual exposure during refueling" because the chemical composition of the two exposures are not comparable; he added that MVMA had asked the Health Effects Institute to assess the health effects of gasoline vapor and explore research needs.

Following these remarks, Chairman Griffin asked for the comments of the invited consultants beginning with Ronald Wyzga, the expert on quantitative risk assessment. Wyzga's "greatest concern" about the staff paper was that uncertainties in the numerical risk estimates were "probably underrepresented." Concurring with Vostal, he emphasized that "the experiment used atomized gasoline whereas humans are exposed to aromatic vapors." He acknowledged that most of the uncertainties could only be resolved through further research and that such research could take a long time to complete.

Samuel Lestz, the fuels consultant, followed by showing that the unleaded gasoline used in the experiment matched closely many of the physical properties of gasoline on the market. He added, however, that "the design of the experiment left something to be desired because the material that was inhaled by the animals was not representative of the material that is, or has been found to be, in the vapors that are in the breathing zone of people who come into contact with gasoline vapor." He felt that the experiment should have been designed to expose animals to the light materials in gasoline (say, from C_4 to C_6).

The principal investigator of the API study, H. N. MacFarland, was present at the meeting and responded to Lestz's comment. He said that the bioassay was a hazard identification study and was never intended to be used for human risk assessment. This was the reason, he explained, "why whole gasoline was vaporized and not some light ends from the gasoline." He concurred that "what people are exposed to in the real world" is "completely different" from what "our animals were exposed to in the chamber." In follow-up research he indicated that API was looking at the role of heavy fractions as a mechanism for renal toxicity in the rat.

The committee then heard comments on "old rat nephropathy" by the principal pathologist for the API study, and by Jack Gray, the committee's consultant on pathology. The theme of this discussion was that the excess tumors in the rat kidney might follow modes of pathogenesis that are not what we see in humans. Trump emphasized that research was in progress to determine whether these rat kidney tumors were a nongenotoxic response, perhaps a secondary effect of renal toxicity.

Marvin Kuschner emphasized that excess liver tumors in female mice were also found in the API study. This data provided support for EPA's qualitative finding of carcinogenicity, even if the rat data were discounted. Trump hypothesized a nongenotoxic, promotional response in the mouse liver but confessed that scientists just did not know the mechanism with any confidence.

The first part of the meeting concluded with a review of the relevant epidemiology by an API consultant, Philip Cole of the University of Alabama. Cole criticized the EPA staff paper for reviewing only three of the eleven epidemiological studies available but concurred with EPA analysts that the overall pattern of epidemiological findings was inconsistent. None of the studies demonstrated a clear dose-response relationship between gasoline and cancer mortality.

Conclusions of Individual EHC Members

In the second part of the meeting, Chairman Griffin moved the discussion to conclusions and recommendations about the EPA staff paper. Leading off, Morton Corn praised the scientific quality and validity of the API's animal bioassays and said that the tumors should serve as an "alert." But Corn insisted that there were many uncertainties about potential human risk and that "it is difficult for me to endorse their [EPA] doing a [risk assessment]." He expressed a fear that "the resulting risk numbers would be widely used and widely quoted" without regard to the complexities and unknowns. Edward Ferrand immediately concurred with Corn's views, both the need for an alert about this "suspect material" and discomfort about "doing a risk assessment on such vague numbers."

Warner North endorsed the comments of his colleagues, particularly the "underrepresentation of uncertainty" in the EPA staff paper. He stressed that "it's really problematic to have a risk assessment such as is described in this report, or at least hinted at, served up as the basis for Agency decision making." Instead of simply discouraging risk assessment, North urged EPA to go beyond a single-number approach and "strive mightily to do justice to the uncertainty and the complexity of this problem." In particular, he suggested that the following complexities be addressed:

- The differences in chemical compositions of exposure between the test animals and humans
- The possibly less important role of microscopic, nonmetastatic tumors in the animals relative to malignancies
- The potential nongenotoxic effects in the kidney, suggesting the possibility of a no-effect threshold or a highly nonlinear dose-response function
- The compatibility of the rodent-based risk estimates with the published epidemiological data

North emphasized that he had "very strong concerns about extrapolating from past [CAG] practice in [risk assessment] for this particular situation."

Michael Symons emphasized another source of uncertainty in risk assessment: the fact that humans face intermittent exposure to gasoline vapor, while the test animals had continuous exposure. EPA analysts should "document as best we can whatever we might know about the effects of giving the body time to recover effectively." While offering some criticism of the staff paper, Symons praised CAG for being informative about uncertainties in the choice of various low-dose extrapolation models and in the reporting of both maximum-likelihood estimates of risk and upper-bound figures. But in the final analysis Symons, noting the tremendous uncertainties, said that "[the risk assessment] shouldn't be used as a basis for a regulatory decision, the computation shouldn't be used for that."

Ronald Hood concurred with Symons and emphasized that "the metabolism of the absorbed material could very well differ considerably in the continuously exposed animal model versus quite intermittently exposed humans in many cases, which could then influence the outcome of such an exposure." He expressed skepticism, however, about whether the epidemiology was of high enough quality to serve as a reliable check on the animal-based risk estimates.

Seymour Abrahamson agreed generally with the comments of Warner North about the need to do a more sophisticated risk assessment. He noted that the risk estimates presented in the staff paper, crude as they were, were not far out of proportion to other occupationally exposed risks. He speculated that "these numbers are probably very, very high for the reasons that have been mentioned yet I don't find these risks to be that alarming."

Daniel Menzel endorsed the practice of quantitative modeling of risk and argued that the real issue concerns how one "lays out the limits of such modeling and how one interprets the uncertainties associated with such models." He thought that the efforts of CAG to convey uncertainty in the staff paper were "a major step in the right direction." On the promise of future research, Menzel expressed skepticism about whether the mechanistic research in progress would shed much light on the modeling questions. He noted: "I see no hope in standing around and waiting for this [mechanistic] data. Therefore, I would urge the EPA not to do that." Since the chemical composition of gasoline is "ever changing" due to economic concerns, he said that no long-term animal study is likely to produce results that are exactly relevant to the gasoline then in use. On "the bottom line of this particular exercise," Menzel agreed with EPA's conclusion that gasoline vapor should be called a 2B carcinogen within the IARC categorization scheme.

Herman Collier endorsed the API study as a sound exercise yet expressed the "hope that the Agency would hear the recommendation that the risk assessment not be applied to the question of human exposure to vapors in gasoline until some further evidence from the other work can be included." Jack Hackney responded similarly: "The uncertainties in the model and in the empirical data are so great that the adequacy of the approach cannot be judged accurately in this case." Therefore, Hackney "doubt[ed] whether its use is appropriate for regulatory policy decisions about public health risk from exposures to gasoline vapors via the gasoline marketing system." Marvin Kuschner seemed to echo the dominant sentiment in the group in his concluding remarks: "So I come out where I guess everybody else comes out, that yes, this material does cause cancer in animals in a way that is consistent with all of the testing that has been witnessed; no, we do not have enough information to base a decision on a human risk assessment."

John Doull expressed a somewhat different view. He saw the compositional differences in gasoline vapors from test animals to humans as "a quantitative rather than a qualitative difference." He insisted that "those are the kinds of things you can sometimes correct for, they're not fatal to a tox study in using that information for prediction." What Doull found troubling were the disease endpoints—the possibility that rat kidney tumors and mouse liver tumors are not predictive in man. He nonetheless supported the designation of gasoline vapor as a 2B carcinogen. On whether risk numbers should be generated, Doull was clearly skeptical, stating that "I think it's important to make decisions that are highly credible, that will be defended by the scientific community and will be logical in terms of the public." He was not precisely sure where he came out but expressed sympathy with the positions of Corn and North.

Bernard Weiss was more explicit:

I agree that the data from the API study implicates gasoline as a possible carcinogen, but I think that issuing a unit risk estimate is premature, and as Dr. Ferrand has indicated, once such an estimate is issued, it's taken up by various state and local authorities as "Biblical" information and used for regulations. I think EPA has to await further information.

The only committee members who elected not to state an overall conclusion were William Schull and committee chair Herschel Griffin, who saw his role as that of a moderator. As the meeting came to a close, Griffin charged Terry Yosie (SAB director) with the task of drafting a report of the committee's deliberations.

Reactions to the Public Meeting

Key staff members of the Motor Vehicle Manufacturers Association and the American Petroleum Institute left the July 25 public meeting with a sense of relief. Fundamental questions had been raised about the validity of doing a quantitative risk assessment to support regulation. A reporter from the trade journal *Automotive News* summarized the public meeting with the headline: "EPA Arm Recommends End to Gas Vapor/Cancer Inquiry." The article explained:

EPA's Science Advisory Board has recommended that an inquiry into the possible cancer-causing effects of breathing gasoline vapor not progress to the formal risk assessment phase.

Although concerned with several studies showing a possible link between cancer in laboratory rodents and their exposure to gasoline, as well as "real world" data on gasoline-handling workers and their rate of cancer, the board decided the data did not support a strong link between gasoline vapor and a significant increase in human cancer.

The science experts' unanimous decision means EPA will probably not enter the risk assessment phase, which ultimately could have meant expensive modifications to vehicles or gasoline-handling equipment to control vapors.[25]

EPA staff drew a quite different message from the public meeting—or at least from the subsequent written report. For example, David Cleverly of OAQPS explained: "SAB supported EPA conclusions that gas vapor is a carcinogen, despite arguments to the contrary. . . . To say gas vapor is a carcinogen implies it should be regulated."[26] EPA scientist Peter Preuss added: "The air office [OAQPS] is required to bring work to SAB for review. The office needs a certain amount of agreement from SAB before it can go forward. Gas vapor got a supportive enough review to go forward. The review was not a veto of the rulemaking process."[27]

Written Report of the EHC

SAB Director Terry Yosie prepared a draft summary of EHC's key conclusions and a draft cover letter to EPA Administrator William Ruckelshaus from EHC Chairman Herschel Griffin and SAB Chairman Norton Nelson. The drafts were distributed to each EHC member by mail, and comments and revisions were requested by Yosie via telephone.[28]

MVMA staff obtained a copy of the drafts and were appalled at what they read. They persuaded Fred Bowditch (vice president of MVMA) to write the chairman of EHC, Herschel Griffin, with specific criticisms of the draft material. Bowditch wrote that "the draft letter and attached comments are misleading because they do not accurately and completely reflect the discussion of certain major issues by Committee members." In particular, the difference in chemical composition of wholly volatilized unleaded gasoline and refueling vapors was emphasized at EHC's public meeting yet "inaccurately expressed" in the draft letter and report. Bowditch was especially irritated by the statement that "the issue of representativeness of the inhaled vapors, while certainly a complicating factor in the study, should not be regarded as a major flaw." On the basis of a review of transcripts from the July meeting, Bowditch insisted that EHC members believed that it was a "major flaw" to apply the results from test exposures to human exposures during refueling.[29]

The final version of the EHC letter and report was submitted to EPA Administrator William Ruckelshaus on October 29, 1984. According to Yosie, the final EHC report and letter contained only slight editorial revisions, those made in response to suggestions by individual EHC members. The letter summarized EHC's major conclusions as follows:

The Committee believes that the [API] study was well designed and that the investigators utilized appropriate scientific protocols that support the reported results.

The Committee agrees with this [EPA's] conclusion [that] wholly vaporized unleaded gasoline should be classified as probably carcinogenic to humans, according to the classification procedures developed by the International Agency for Research on Cancer.

The Environmental Health Committee believes that the [EPA] analysis under-represents the degree of uncertainty in assessing human health impacts from this complex mixture of pollutants.[30]

The accompanying report contained a more detailed discussion of these three conclusions—highlighting in the end the numerous sources of uncertainties in the risk estimates. In conclusion, the report stated: "The EPA may not be able to resolve all of these issues based on the information available at this time. However, the Committee wishes to raise them for the purposes of comprehensiveness."[31]

EPA staff responded to EHC's letter and report in a letter to Herschel Griffin from EPA Administrator William Ruckelshaus. While volunteering to make some minor changes in response to EHC's comments, the letter stated that the compositional differences in gasoline vapors were "acknowledged in the Staff Paper to be an uncertainty and remains as one of the major qualitative uncertainties in our analysis." The letter did promise some further analysis of the benign tumors but expressed the view on the other issues that EPA "sees no alternative other than to use the standard assumptions common to other risk assessments where the data are lacking."[32]

API'S STRATEGY

The SAB review process only served to intensify the strategy that API had been following since 1983: discover the mechanisms of the kidney tumors and stay abreast of worldwide scientific developments. In July 1983, for example, API sponsored a scientific meeting in Boston where two informal working groups, one on epidemiology and one on toxicology, charted a research agenda for the future. It was at this meeting that Carl Aldren of Proctor and Gamble advanced the hypothesis that the kidney tumors in the male rat are produced by a mechanism unique to the male rat. In the absence of proof that something like this was true, industry scientists were convinced that "EPA would hold us to the data in the most sensitive species when doing human risk assessment."[33]

The petroleum industry was no novice at waging war with Washington on chronic health and regulatory issues. While it lost the lead issue to EPA in the 1970s, the industry got smarter about how to play the regulation game. In fact, the gasoline vapor controversy emerged on the heels of API's big victory at the Supreme Court in the OSHA benzene case. In light of their prior experience, API scientists knew that their best bet was to persuade EPA through good science that gasoline vapor did not pose a significant risk of cancer to humans.

API scientists believed that the key scientist at the Ruckelshaus EPA was Bernard Goldstein, the assistant administrator for research and development. Industrial scientists were not sure where Goldstein would come down on a given issue, but his academic reputation was excellent.[34] Goldstein was clearly his own man and seemed to have a strong public health orientation.

API scientists felt that they were "victimized" by the SAB process.[35] While the public hearing was fair and the committee discussion encouraging, the letter to the administrator was perceived by industry to be a misrepresentation of that discussion. Many industry observers suspected that SAB's Executive Committee rewrote the SAB's final report, deemphasizing the uncertainties, to facilitate regulation.[36]

A review of the transcript does in fact reveal discrepancies between the tone of the public hearing and the tone of the final SAB report to the administrator. According to Terry Yosie, some differences should be expected, since there was more to writing the final report than simply summarizing the transcript. Due to the brevity of the report, certain ideas and opinions expressed in the public meeting would be better represented than others; as a result, the distribution of opinions would be different in the report than in the meeting. Yosie noted, however, that everyone had ample opportunity to object to the contents of the report when it was circulated. In general, if concurrence of all committee members is not reached when a summary report is circulated, a minority report can be written. No minority report accompanied the gasoline vapor final report.

By the end of the SAB process it was apparent that only scientific progress could resolve the industry's predicament on gasoline vapor. A concerted, multimillion-dollar research program on the mechanisms of gasoline toxicity was urgently needed. While API could find the resources to finance such a program, it did not have direct access to the necessary scientific talent.

One option was to work through a reputable university-based laboratory. However, API had recently had several frustrating experiences with university-based laboratories. Reports were hopelessly behind schedule, and API felt that it could not rely upon such a process.[37]

API scientists decided to approach James Gibson, vice president of research for the Chemical Industry Institute of Toxicology in Research Triangle Park, North Carolina. CIIT was a well-respected industry-supported research organization. Gibson explained, however, that CIIT was not interested in mixtures and therefore denied the request. Since many members of API were also financial supporters of CIIT, a second attempt was made during CIIT's annual request for nominations of research topics. A breakthrough occurred when CIIT board members James Mathis (Exxon), Paul Deisler (Shell), and James McCullough (Mobil) worked together and made the case to CIIT's leadership.[38]

A potential problem arose when it became apparent to API that no CIIT data could be discussed publicly until it was submitted for publication. CIIT's policy of nondisclosure was based on a sound rationale: the need to prevent one company from gaining competitive advantage over another through access to new scientific data. CIIT's counsel, Milton Wessell, worked around this problem by creating a working group of industry people that could monitor progress on gasoline vapor without compromising CIIT's policy of secrecy about data.[39]

The final result was a $4-million CIIT research program on gasoline, half financed by CIIT and half by API. Two of CIIT's best scientists, James Buss and James Swenberg, were recruited to explore a series of scientific questions critical to the future of the petroleum industry. As this program of original research was being launched at CIIT, a very different strategy was being devised by the other major industry that was threatened by EPA rulemaking.

MVMA'S STRATEGY

The Motor Vehicle Manufacturers Association is the industry association that represents the domestic motor vehicle manufacturing companies. Fearing that the EPA staff paper was analytical ammunition for on-board controls, MVMA staff saw the SAB review as a critical hurdle for proregulation forces at EPA. As mentioned earlier, MVMA's principal position on the science was that the test exposures in the API study were not applicable to the gasoline vapors inhaled by humans throughout the gasoline marketing system.

The July 25 public meeting was the first time an issue critical to MVMA had been reviewed by the Environmental Health Committee. Before EHC met, MVMA staff were worried about what would happen, according to MVMA's director of technical affairs, Lawrence Slimak:

Based on our prior dealings with other units of SAB, we didn't trust the process going in. We saw the SAB staff as EPA staff, who have a regulatory mission. While the scientific membership of SAB is balanced, some of the key scientists are very political. We didn't know what to expect from EHC, but we knew the regulatory train at EPA was moving.[40]

Slimak decided that MVMA needed to take a proactive position on the issue of health effects from refueling issues. Since MVMA lacked the resources to engage in an original research program, Slimak pursued a different course. He persuaded MVMA's Environmental Health Advisory Committee—a group of technical people from member companies—to request an independent scientific review by the Health Effects Institute. Since HEI was funded by EPA and the major member companies of MVMA, Slimak was confident that HEI would seriously consider this request.[41]

MVMA's request for HEI review was submitted in writing to HEI.[42] At the same time, Slimak wrote two assistant administrators at EPA, Joseph Cannon (Air and Radiation) and Bernard Goldstein (Research and Development) and requested that they join in requesting that HEI review the scientific evidence on gasoline vapor and health effects.[43]

Slimak's letters, dated July 19, were mailed prior to EHC's public meeting on July 25. Slimak acknowledged that the HEI gambit might be viewed by others as "an end run around the SAB." He explained his rationale:

We have much more confidence in the HEI process because it is less entangled in the political game than SAB. We were confident that an objective scientific review would support our position. In any case, we felt we had little to lose by going to HEI since career EPA officials seemed determined to regulate us.[44]

Slimak emphasized that HEI's involvement would be "fair game." Indeed, he felt that HEI had been created in part to fill precisely this kind of role. While

HEI's main mission was to generate original scientific information, Slimak felt that "paper reviews" were clearly authorized by HEI's charter.[45]

MVMA staff were bitterly disappointed about the tone of the final report of EHC, which was released in late October 1984. While Slimak and his colleagues had been encouraged by the July 25 public meeting, they felt that the final report of EHC created a "misimpression" through the "selective presentation" of the facts and views.[46] According to MVMA's health scientist, Richard Paul, "The EHC's letter to Administrator Ruckelshaus misrepresented the extent of the Committee's skepticism and the validity of EPA's quantitative risk estimates."[47] As calendar year 1984 came to a close, MVMA saw HEI as one of its few remaining hopes on the health effects of refueling.

SCIENTIFIC REPORT OF THE HEALTH EFFECTS INSTITUTE

The Health Effects Institute (HEI) is a nonprofit research organization located in Cambridge, Massachusetts, that is jointly and equally funded by the U.S. Environmental Protection Agency and motor vehicle and engine manufacturers. HEI arose out of the adversarial struggles and mutual mistrust that characterized EPA's relationship with the auto industry in the 1970s. The function of HEI is to provide a scientific foundation for regulation of motor vehicle emissions that both EPA and the auto industry can trust.

Although HEI's primary mission is to fund scientific research about the health effects of motor vehicle emissions, the charter of HEI does authorize the organization to assess and evaluate information generated by non-HEI research programs. The creators of HEI recognized that on occasion EPA and the motor vehicle industry would turn to HEI as a conflict-resolution device when disputes emerged about what science says about the health effects of emissions.

HEI was created in 1980, but the first research projects were not under way until mid–1983. The organization was just beginning to operate at full capacity when the MVMA requested that HEI consider a research initiative on the health effects of gasoline vapors. Indeed, the MVMA request was the first time that HEI received a request from a sponsor to undertake a scientific review on an issue of imminent regulatory significance.

HEI's Response to MVMA's Request

When HEI received MVMA's request in July 1984, the organization had just undergone a transition in leadership. The founding executive director, Charles Powers, was succeeded by Thomas Grumbly, who confronted the gasoline vapor issue with roughly two weeks of experience at HEI. Grumbly felt that MVMA's request raised "legitimate scientific questions" that were an important source of disagreement between EPA and segments of the auto industry.[48] Grumbly also saw the gasoline vapor issue as an opportunity and challenge for this young organization to demonstrate its credibility in a live regulatory controversy. As a prelude to granting this request, Grumbly sought advice from Bernard Goldstein, EPA's assistant administrator for research and development, and approval from his Board of Directors (William Baker, Archibald Cox, and Donald Kennedy).

The HEI board approved Grumbly's recommendation that HEI should grant MVMA's request. According to Stan Blacker, HEI's coordinator at EPA, the go-ahead decision was "a high-risk strategy."[49] OAQPS and OMS officials were determined to control refueling emissions and HEI's involvement might be perceived by them as a blocking tactic. Moreover, Goldstein of EPA soon informed Grumbly in writing that the agency could not assess its need for HEI involvement until after the SAB review was completed—which did not occur until late October 1984. Thus HEI embarked on a review of the carcinogenicity of gasoline vapor just prior to completion of SAB review and just prior to a widely anticipated EPA rulemaking on refueling emissions, without the formal blessing of senior EPA officials.

Knowing that EPA's blessing would strengthen HEI's case for involvement, Grumbly worked with Stan Blacker of EPA to include a gasoline vapor review on EPA's annual statement of research needs for HEI. The issue was included as a lower-priority item on a draft needs statement prepared by Blacker. Surprisingly, this priority was not contested by any of the major offices within EPA, including OMS, OAQPS, and OHEA. Blacker remarked that "gas vapor could easily have been removed from the list if any unit within EPA had raised a serious objection." "Few EPA officials realized," said Blacker, "how sophisticated a review HEI would be capable of producing."[50]

In February 1985 Goldstein of EPA submitted the agency's formal needs statement to HEI. The inclusion of gasoline vapor as a lower-priority item meant that both of HEI's sponsors were seeking the scientific review that was already well under way.

The HEI Process

Grumbly had begun HEI's involvement by dispatching one of his senior scientific staff, Robert Kavet, to SAB's public hearing on gasoline vapor in July. Kavet's detailed report to Grumbly on the hearing concluded: "I feel that we must stay abreast of all scientific developments and further develop our planning for a range of situations that could develop. A discussion within staff, followed by presentation of the issue to the Health Research Committee, would be helpful toward defining our role."[51]

The role of the HEI's Research Committee, which is comprised of roughly a dozen academic scientists, is to chart HEI's priorities and open new topics for original investigation. At a September 1984 meeting the Research Committee concluded that it could not intelligently embark on a research agenda for HEI without a comprehensive review of the available data on gasoline vapor. This task was ultimately assigned to HEI's other operating committee (the Health Review Committee) in consultation with Kavet.

To assist in the data-assessment process, HEI commissioned two reviews of the literature: one by Battelle Northwest Laboratories and the other by Environ Corporation. The Research Committee chose to commission two competing reviews because it did not want to be strongly influenced by any one contractor. The work was restricted to the toxicology of gasoline vapors since a scholarly review of the epidemiology was already available. The reports were commissioned in late January 1985 and received by HEI on May 1, 1985.[52]

As HEI launched into its review of gasoline vapor, the potential regulatory implications of HEI's participation became increasingly evident. Grumbly wrote the HEI's Board of Directors on April 23, 1985:

There is some belief in both EPA and the industry that HEI wants EPA to wait on the regulatory front on this issue until HEI completes a *research program* in the area. This would take 2–4 years. I don't believe it is in our interest to take this position, as it exposes us to criticism that we are blocking public action in an area that affects nearly every American.[53]

Senior EPA officials, armed with a "passing" report from SAB, were eager to initiate a rulemaking on gasoline vapor, with cancer risk as a prominent rationale. Grumbly urged EPA officials to wait only through the summer until HEI's review and research agenda were published.[54]

In the spring of 1985 HEI's research and review committees held a joint workshop to review the work of Battelle and Environ. By this time an ad hoc group of HEI scientists with special interest in gasoline vapor had formed. This group included Arthur Upton, Sheldon Murphy, Gerald Wogan, Robert Levy, and Roger McClellan. Upton, an eminent cancer scientist with prior government experience, agreed to be the primary reviewer of the upcoming HEI report on gasoline vapor. At the spring workshop this group also heard a review of the epidemiological data by Philip Enterline, a consultant to MVMA. The results of the workshop were informally communicated to MVMA and EPA officials in early summer of 1985. At this point it was becoming clear to everyone that HEI's opinion about the quality of evidence on cancer risk would be more cautious than the view expressed in EPA's staff paper and approved by SAB's Environmental Health Committee.

During the summer Upton supervised production of a written evaluation of the data base and a research agenda. This document was revised in response to comments from members of the Research Committee, Review Committee, and Board of Directors. The key staff work was executed by Kavet with selected assistance from Grumbly and Ken Sexton.

Findings of HEI's Review

In September 1985 HEI released its report, entitled *Gasoline Vapor Exposure and Human Cancer: Evaluation of Existing Scientific Information and Recommendations for Future Research.*[55] The report highlighted two primary sources of uncertainty associated with using API's animal data as a basis for human risk assessment. Each major source of uncertainty was accompanied by a series of specific research recommendations that might reduce uncertainty.

First, the gasoline in the API animal study was "entirely vaporized for animal exposure, meaning that the inhaled mixture was identical to that in the liquid phase."[56] HEI emphasized that "this mixture is not representative of the evaporative mix found in ambient situations because of the differential volatility of the hydrocarbon compounds present in gasoline."[57] The higher-molecular-weight compounds—those that are present in lower proportions in ambient vapors than in those generated in the API study—"appear most likely to be responsible for toxic effects in the kidneys (nephrotoxicity) of male rats exposed to wholly vaporized gasoline."[58] HEI noted in its research recommendations that "a chronic animal study that uses test atmospheres representative of ambient human environments could lessen uncertainties concerning toxicity and carcinogenicity."[59]

Second, HEI found that "problems remain concerning the applicability of the experimental animal models to humans."[60] The B6C3F1 mouse strain used in the API study has a high spontaneous incidence of liver tumors, and this incidence can be enlarged by application of nongenotoxic chemicals. It is possible, therefore, that the increased number of liver tumors in mice is minimally relevant to humans. Furthermore, the male rat has a "high susceptibility to hydrocarbon-induced renal toxicity," and thus API's results in male rats may have diminished relevance to humans.[61] HEI recommended that research be undertaken to elucidate the mechanisms of hydrocarbon-related tumor formation in male rats and female mice.[62]

In addition to these two major uncertainties, HEI noted that information on human exposure to gasoline vapors was limited and that the published epidemiological studies lacked sound exposure information. More work could be done, HEI noted, to "characterize exposure patterns for various high exposure groups" and to use "industrial hygiene techniques for developing retrospective exposure histories."[63]

In the conclusion of the report, HEI acknowledged that wholly vaporized gasoline is "an animal carcinogen and a presumptive human carcinogen" according to IARC criteria.[64] Yet humans are not typically exposed to wholly vaporized gasoline during the course of their daily activities. Moreover, the suitability of the animal models for human risk assessment is "questionable."[65] In light of these uncertainties, HEI concluded that "development of a meaningful and realistic quantitative risk assessment is very problematic."[66]

HEI emphasized that "actual risk [to humans] from ambient gasoline vapors might be anywhere between zero and the upper bound [reported by EPA]."[67] The report expressed no explicit policy opinion about whether regulatory action should be implemented immediately or delayed until the results of further research were available. If EPA should determine "on the basis of policy that further research is needed prior to or concurrent with regulatory action, then specific projects may be developed by HEI and others from the recommendations presented above."[68]

It should be noted that none of these issues was new to the gasoline vapor controversy. The HEI report emphasized points that had been raised by speakers at the SAB public hearing and, to some extent, in the EPA staff paper itself.

REACTIONS TO THE HEI REPORT

The HEI report became a hot document in Washington because EPA was widely perceived as about ready to propose a national on-board control program. Inside the agency the cancer threat from gasoline vapors had been advanced as an important rationale in the case for on-board controls. Career EPA officials in OHEA, OAQPS, and OMS were, not surprisingly, irritated by the conclusions in the HEI report. HEI, said one EPA official, "threw a monkey wrench into the proceedings."[69]

According to Terry Yosie of SAB, many career EPA officials believed that HEI had "overstepped" its role as a scientific research organization.[70] It was not HEI's role, they argued, to tell EPA when to do a quantitative risk assessment. Some observers felt that HEI should never have produced such a report because the organization's primary role is to produce new knowledge. David Cleverly of EPA explained: "If HEI wants to do policy, they should be open about it. They shouldn't pretend to do hard core science. In this case they were surreptitiously involved in policy without acknowledging it."[71] Roger McClellan of HEI's Research Committee insisted that this was not HEI's intention: "HEI struggled hard to be objective. We didn't want to foreclose regulatory options."[72]

Other EPA and HEI officials felt that it was appropriate for HEI to offer reviews of the science on key regulatory issues, but the gasoline vapor review came too late in the process to be constructive. If HEI had intervened earlier when EPA's stance on cancer risk was more fluid, the review could have been quite influential.[73] Once EPA's position on cancer risk was formulated and approved by SAB, HEI's intervention was destined to be less influential and more controversial.

A minority view among HEI officials was that the HEI report was actually helpful to proregulation forces because it did not completely discredit EPA's position on cancer risk. Elizabeth Anderson of OHEA explained: "The HEI report had a steadying effect. By not throwing stones at EPA's position, HEI bridged the gap between API's and EPA's views. It prevented a situation with EPA in one corner and API in the other."[74] Others emphasized that HEI was simply offering a pure science position by pointing to gaps in the data. Having such a perspective considered in the rulemaking process might increase the credibility of the agency's final decision, whatever that decision proved to be.

Some observers felt that EPA officials had put themselves in a precarious position by placing too little emphasis on ozone control and too much emphasis on the speculative cancer threats. Consultant Michael Walsh explained that the HEI report should never have had a big impact on the regulatory process because the cancer issue "was only one of many factors operating in favor of an on-board control program."[75] Bernard Goldstein made the same point somewhat differently: "The impression I and others had was that the real goal of those in EPA pursuing gasoline marketing controls was the need to reduce ambient ozone levels, and that the cancer issue was a convenient stalking horse."[76]

While MVMA staff were generally pleased about the HEI report, API staff were somewhat uneasy. There were suspicions that the HEI report was intended to discourage EPA from requiring on-board controls.[77] If ozone was the only concern, EPA might be more inclined to hit the petroleum industry with a Stage II regulation. If this was the case, API's vigorous research program at CIIT on the mechanisms of kidney toxicity was beginning to exert the same kind of policy influence.

HEI'S RESEARCH STRATEGY

After publishing its report on the gaps in the science on gasoline vapor and cancer risk, HEI faced the question of whether to launch its own original research program in this area. HEI's key staff scientist, Robert Kavet, held the view that HEI should play a role in resolving some of the uncertainties that had been uncovered. In particular, he recommended to HEI's Research Committee that a long term animal bioassay of partially volatilized gasoline vapors (similar in composition to those inhaled by humans) be undertaken.[78] There was no significant interest at API, MVMA, or EPA in funding such a study.

In order to clarify what original research was already under way, HEI sponsored a "Gasoline Vapors Research Workshop" in Cambridge in December 1985. It received a thorough briefing on the research programs at API, CIIT, EPA, and several petroleum companies and vehicle manufacturers. The prime objective of this workshop was to identify research activities that would, if undertaken, narrow critical gaps and thereby assist HEI's Research Committee in charting its future funding priorities.

After the workshop and upon careful deliberation, HEI's Research Committee decided against launching an original research program on gasoline vapors. Two rationales were offered for this decision.[79] First, the critical mechanistic work on the male rat kidney was already well under way at CIIT under API sponsorship. Second, a new HEI research program might be perceived by EPA as an effort to block the forthcoming rule on refueling emissions, even if that would not be HEI's intention. Given the work in progress at CIIT and the political sensitivity associated with a major research program, HEI elected not to pursue original research in December 1985.

The wisdom of HEI's choice is disputable. Given that some felt that HEI was overstepping its bounds as a research organization by conducting a review of existing data, initiating basic research as a result of the review might have served to increase HEI's credibility, at least in the eyes of some career EPA officials. At least HEI would have been seen as making a genuine effort to resolve the gaps in knowledge that it was so eager to highlight.

CIIT'S MECHANISTIC RESEARCH

Bernard Goldstein left the EPA in early August 1985 and returned to academia. As a premier cancer scientist, he kept abreast of new scientific discoveries about carcinogenesis and was especially interested in CIIT's ongoing work on the causes of kidney tumors in male rats. In the fall of 1985 some preliminary data were emerging from CIIT that were quite relevant to EPA's forthcoming decision on gasoline refueling emissions.

CIIT's basic scientific research into the mechanism of gasoline-induced kidney tumors was suggesting that the tumors were dependent upon the presence of a specific low-molecular-weight protein produced under hormonal control by male rats. This protein, alpha–2-microglobulin, is excreted in relatively large concentrations into the urine and then reabsorbed at the kidney tubule anatomic site at which the tumor is formed. The protein appears to facilitate the entry and/or action of the particular gasoline hydrocarbons responsible for toxic effects. These kidney effects are not observed in female rats or in castrated male rats. A similar protein is not known to occur in humans, although this possibility cannot be completely ruled out, at least at low concentrations.

Goldstein was sufficiently impressed with these developing findings that he telephoned and personally told Lee Thomas (Ruckelshaus's successor at the helm of EPA) that "there is a reasonable possibility that the kidney tumors in male rats will turn out to be due to a specific mechanism not pertinent to humans." According to Goldstein, this was one of only a few occasions after his departure from the agency that he initiated a phone call to Thomas on the basis of new science that was pertinent to his decision process.[80]

PERSPECTIVE OF EPA ADMINISTRATOR LEE THOMAS

Lee Thomas succeeded William Ruckelshaus as administrator of EPA in February 1985. Refueling emissions became a live issue early in Thomas's tenure because the Office of Mobile Sources was seeking approval of an ambitious rulemaking proposal that called for mandatory on-board controls. Although Thomas had worked for EPA's Office of Solid Waste and Emergency Response for several years before succeeding Ruckelshaus, he was a relative novice about the issues of quantitative cancer risk assessment. Thomas explained that "risk assessment was still fairly new as an across-the-agency activity when I took over." Since the Superfund and solid waste programs did not routinely use risk assessment, Thomas "didn't have much experience with the assumptions and data used in cancer risk assessment."[81]

By the summer of 1985 Thomas had been briefed on the refueling issue and was tentatively ready to sign off on a proposal for on-board controls. Cancer risk had been advanced as a prominent rationale for regulation in addition to the agency's historical interest in ozone control. Thomas recalled being briefed on the SAB's review of the EPA staff paper and being told that "SAB basically agreed with our staff." The key SAB finding in Thomas's mind was the determination that gasoline vapor is a "probable human carcinogen."

Late in the summer Thomas was informed by special assistant Deb Taylor that the Health Effects Institute had looked into gasoline vapor and was convinced that "the toxicological data and the risk numbers are not as solid as EPA staff had previously represented them to be." Thomas had not been aware of the HEI review because it had been initiated during the tenure of Ruckelshaus. Taylor told Thomas that the HEI report was about to be published.

Thomas immediately asked for a briefing on gasoline vapor by the Office of Health and Environmental Assessment. The briefing was delivered by Elizabeth Anderson, head of OHEA's Carcinogen Assessment Group. Thomas recalled that her briefing contained "no serious discussion of the uncertainties." Her theme was that "the data on gasoline vapor are no softer than what we are used to in the cancer area." Thomas recalled "being impressed about how little equivocation" there was in the CAG presentation on gasoline vapor. Thomas also had a vivid memory of the thought: "If this is a garden variety case, most of our cancer decisions are flimsy."

When the HEI report arrived at EPA, Thomas told Terry Yosie (SAB) that he wanted to hear a scientific dialogue on the toxicity of gasoline vapor. It was agreed that the participants would include representatives from EPA, HEI, and SAB. The so-called Briefing for the Administrator was conducted on November 26, 1985. It began with presentations by Robert McGaughy (OHEA-CAG) and Arthur Upton (HEI) and evolved into an open dialogue on the scientific issues, including specific questions by Thomas. The HEI participants included Upton, Gerald Wogan, and Thomas Grumbly. SAB was represented by Terry Yosie and the new chairman of the Environmental Health Committee, Richard Griesemer. Numerous EPA officials were present from OMS, OHEA, OAQPS, and ORD.

Thomas described the experience as follows: "The HEI meeting gave me a good hit with the uncertainties." From the dialogue Thomas learned that real-life human exposures to gasoline vapor were not comparable to the test-animal exposures in the API study in terms of the amount of "heavy-end" hydrocarbon compounds ($\geq C_6$). He also learned that the agency did not know precisely what the chemical composition of inhaled vapors was in most real-life settings. Since the $\geq C_6$ fractions were probably responsible for toxicity, this information was crucial. Thomas recalled that "I had not heard this from my staff presentations." The HEI meeting also raised uncertainties in Thomas's mind about the relevance of the rat kidney tumors to human risk assessment, especially in light of preliminary data from CIIT. Thomas remembered thinking that these issues were "big uncertainties."

Terry Yosie disagreed with Thomas's assertion that he was not thoroughly apprised of the uncertainty issues before the HEI meeting. Yosie felt that Thomas received "plenty of signals" regarding the uncertainties, but because of his inexperience with cancer risk assessment, he may have failed to "interpret the signals correctly."[82]

After the HEI meeting, Thomas told his staff to redo the quantitative risk assessment. He wanted to see the numbers redone so that the risks were based solely on the $\geq C_6$ fractions. As Thomas suspected, the modifications resulted in "a big difference in the risk estimates." Thomas instructed his staff that future decision documents could include cancer risk estimates based on benzene and the $\geq C_6$ fractions but nothing more. Soon after the HEI meeting, Thomas concluded that "I didn't have the evidence to go on-board nationwide on the basis of cancer risk." He then was compelled to look more carefully at the ozone nonattainment issues.

The HEI report complicated Thomas's decision process because he had originally thought that he was ready to make a decision soon on refueling emissions. Under pressure from state air officials, who were facing legislated ozone compliance dates, Thomas had promised in the fall of 1985 to make a rulemaking decision on refueling by the end of the calendar year. However, Thomas said that he did not want to ignore the HEI report, especially since it was requested by EPA and was prepared by "a credible independent group with no ax to grind."

The HEI report, in combination with the agency's renewed interest in a comprehensive ozone control strategy, caused Thomas to make his ultimate decision in July 1987 instead of January 1986. The eighteen-month delay was consumed collecting exposure information, redoing the quantitative cancer risk assessments, and making the analysis of ozone control more prominent and sophisticated. The delay was frustrating for Thomas because he "took heat" for violating his promise to the state air officials and was grilled in an interview by Philip Shabecoff of the *New York Times*.

In the final analysis Thomas decided to make the same decision in July 1987 that he had tentatively made in January 1986, but with a new rationale. Cancer risk was no longer the key concern. He concluded that the "on-board systems and Stage II controls in nonattainment areas were roughly a wash from the standpoint of overall ozone control." He leaned toward on-board systems as (1) "an insurance policy against cancer risk" in attainment areas, (2) a protection against transport of hydrocarbons from attainment to nonattainment areas, and (3) an easier strategy to implement and enforce. In effect, HEI's scientific advice made the decision "a much closer call" and caused Thomas "to make the decision on correct grounds."

The gasoline vapor experience "alerted [Thomas] to the fact that the agency needs to spend more time showing the uncertainties in risk assessment." After the HEI advice on gasoline vapor and several other decisions on cancer issues, Thomas created the Risk Assessment Council and the Risk Assessment Forum at EPA to give uncertainty analysis more attention. He also initiated a research program on how to convey uncertainties in quantitative risk assessment. Although the agency's cancer-assessment guidelines were redone once during Thomas's tenure, he has recently made a "tentative decision" to reopen them again to foster more attention to expression of uncertainty.

EPA'S REVISED RISK ASSESSMENT AND REGULATORY IMPACT ANALYSIS

In April 1987 EPA's Carcinogen Assessment Group released a revision to its 1984 staff paper entitled *Evaluation of the Carcinogenicity of Unleaded Gasoline*.[83] The report was intended to update the 1984 report, taking into account new scientific data, public comments, the HEI report, and the SAB review. In July 1987 EPA also released a revised regulatory impact analysis (RIA).[84] The two documents constituted the agency's analytical rationale for the administrator's decision to mandate on-board control systems to reduce refueling emissions.

The 1987 risk-assessment report continued to give substantial credence to the cancer threat, although some changes were made. The key findings were the following:

- There was sufficient evidence to conclude that gasoline vapors are carcinogenic in animals, but the human evidence was inadequate.

- The male rat kidney tumors were not disregarded, despite CIIT's mechanistic research on alpha–2–microglobulin, because "the link between hydrocarbon nephropathy and tumor induction is not proven."[85]

- The plausible upper limit for the increased cancer risk from lifetime exposure to 1 ppm of unleaded gasoline vapor was 3.5×10^{-3}, which represented no significant change form the 1984 staff paper.[86]

The 1987 RIA contained a more significant change in quantitative risk assessment that seemed to be responsive to SAB's and/or HEI's comments. A possible reduction in the annual cancer incidence estimates (from 68 to 17) was made to reflect the fact that "the ratio of C_6–C_9 compounds in the liquid to C_6–C_9 compounds in the vapors is a factor of 4 to 1."[87] The C_6 through C_9 fractions are those thought more likely to be responsible for carcinogenicity. The adjustment in the incidence estimate accounted for the chemical differences between the wholly volatilized gasoline administered to rodents in the API study and the gasoline vapors breathed in daily life by humans. Both the original estimates and reduced estimates of cancer risk were reported in the 1987 RIA.

In the August 1987 *Federal Register* notice of the proposed rule covering on-board controls, the agency's overall position on cancer risk was explained. The two key points were the following:

EPA has concluded that gasoline vapors are a probable human carcinogen under EPA's Cancer Risk Assessment Guidelines.[88]

EPA believes the evidence is sufficient to justify a quantitative estimate of human risk from exposures to gasoline vapors in order to gauge the magnitude of the potential public health problem.[89]

On the first point EPA emphasized ''that ''this conclusion is shared by SAB and HEI.''[90] The second point was accompanied by the following statement:

In reviewing EPA's risk assessment, SAB concluded that the degree of scientific uncertainty in the resulting quantitative estimate of human risk should be clearly acknowledged. HEI expressed a more cautious view regarding the actual quantification of carcinogenic risk, stating that it is not presently possible to draw accurate conclusions concerning the degree of human risk.[91]

Referring to CIIT's mechanistic research on the male rat kidney, the rule-making rationale stated that ''the link between acute or chronic hydrocarbon-induced kidney damage and carcinogenicity is not clear, nor has it been proven.''[92] The agency added that ''it is important to note that carcinogenic response to gasoline vapors was also demonstrated in the API laboratory studies of female mice.''[93]

Finally, in discussing the adjusted risk estimates for gasoline vapors based on the C_6 through C_9 compounds, the rulemaking rationale acknowledged that the unadjusted risk estimates may be too high by a factor of four. However, the agency quoted SAB as follows: ''The issue of the representativeness of the inhaled vapors, while certainly a complicating factor in the bioassay, should not be regarded as a major flaw.''[94] Moreover, the agency ''could find no data to indicate which fraction of the gasoline vapor is responsible for the induction of liver tumors in female mice.''[95] As a result, EPA concluded that ''it would be unwise to base the quantitative risk assessment on an assumption that may significantly underestimate the potential health problem.''[96]

CONCLUSION

At this writing the outcome of the gasoline vapor controversy is still not resolved. The decision by Thomas was bottled up at OMB until the Bush administration took office in January 1989. Congress is now considering the issue in its efforts to rewrite the Clean Air Act. Although the final outcome is unresolved, some important conclusions can still be drawn.

In the final analysis, the potential cancer risk posed by gasoline vapor does not appear to have been a central consideration in EPA's decision to regulate refueling emissions. Concern about ozone nonattainment was the driving force from the beginning and carried the day in the end. However, the cancer threat did play a significant role in Administrator Thomas's decision to choose a national on-board control strategy instead of Stage II controls in ozone nonattainment areas.

The cumulative effect of SAB review, HEI's report, and CIIT's mechanistic research on EPA's cancer risk assessment was quite limited. The agency's qualitative description of gasoline vapor was unchanged ("probable human carcinogen"), while the population risk estimate was changed from a single number ("plausible upper bound") to a numerical range that reflected the possibility that only C_6 through C_9 compounds are responsible for toxicity and carcinogenicity.

SAB review alone appeared to have very little impact on the agency's rulemaking process. Career officials seemed determined to regulate, and the written report to the administrator from SAB was perceived by all parties as an implicit approval of rulemaking. A curious discrepancy exists between the publicly stated views of members of SAB's Environmental Health Committee and the written report of SAB to the administrator, a discrepancy that angered industry observers. Explicit closure of the committee's position at the public meeting would have been preferable. The case underscores why SAB is perceived as an "approval" mechanism in the regulatory process.

The HEI report and the November 1985 meeting with Thomas appeared to have a powerful effect on the rulemaking process, even though the final policy outcome was not changed. HEI informed Thomas of uncertainties that he felt CAG had not thoroughly explained to him in personal briefings. The proposed rulemaking was delayed eighteen months, at least in part due to HEI's intervention. New exposure data were collected, the risk assessment was updated, and the regulatory impact analysis was modified. The critical substantive change in Thomas's mind, which may be attributed in part to HEI intervention, was an increased emphasis on ozone control relative to cancer risk in the ultimate decision rationale, a position that Thomas felt was more defensible.

The gasoline vapor controversy raises fundamental questions about the breadth of HEI's mission. A restricted view is that HEI should focus exclusively on producing new scientific data and avoid any mediating role in sensitive regulatory controversies when interpretation of existing data is at issue. A more flexible view, which is compatible with HEI's charter, would permit HEI to offer opinions on scientific questions with regulatory implications, at least when asked to do so by one or both of its sponsors. Insofar as HEI-supported data become a focal point of controversy in future rulemakings, it may be impossible for HEI-affiliated scientists to refrain from expressing opinions on how their data should be interpreted by regulators.

Clearly, by choosing to conduct a review whose results were reported directly to the administrator, HEI did little to endear itself to the career EPA officials who had worked on the gasoline vapor rulemaking. Should HEI decide to become involved in other regulatory controversies, a more effective strategy may be to become involved earlier in the process, before EPA has committed substantial time and energy to one particular course of action, and to cultivate more positive relationships with career EPA staff.

CIIT's mechanistic research on the male rat kidney appears to have influenced the beliefs of HEI scientists Goldstein and Thomas. However, the agency's official documents consistently downplayed the definitiveness of this research.

HEI's decision not to pursue original research on gasoline vapor was made because CIIT and API were already involved and because HEI did not want to be perceived as blocking EPA's rulemaking ambitions. HEI may have missed the scientific opportunity to sponsor a critical unperformed study: a chronic animal bioassay with gasoline vapors similar in chemical composition to those inhaled by humans, possibly with an intermittent exposure regimen. Negative results in such a study would have been particularly enlightening because acute studies have shown that the light ends of gasoline vapor do not cause nephro- toxicity in rats. In light of HEI's decision to criticize EPA for quantifying risk on the basis of poor data, the decision not to generate better data is questionable. In fairness to HEI, such a research program would have consumed three to five years and would probably have been too late to influence Thomas's decision on refueling emissions.

APPENDIX A: INTERVIEWEES

Environmental Protection Agency

Elizabeth Anderson

Larry Anderson

Scott Baker

Stan Blacker

David Cleverly

Gene Durman

Bernard Goldstein

Robert McGaughy

Peter Preuss

George Sugiyama

Joe Summers

Lee Thomas

James Weigold

Richard Wilson

Health Effects Institute

John Bailar

Thomas Grumbly

Robert Kavet

Robert Levy

Roger McClelland

Ken Sexton

Arthur Upton

Science Advisory Board

John Doull

Herschel Griffin

Marvin Kuschner

James Menzel

Terry Yosie

American Petroleum Institute

Gary Martin

Robert Scala

Steven Swanson

James Vail

Motor Vehicle Manufacturers Association

Richard Paul

Lawrence Slimak

Miscellaneous

David Doniger

David Finnegan

Michael Walsh

APPENDIX B: MEMBERSHIP OF SAB'S ENVIRONMENTAL HEALTH COMMITTEE (JULY 1984)

Chairman

Dr. Herschel E. Griffin
Professor of Epidemiology
Graduate School of Public Health
San Diego State University

Members

Dr. Herman E. Collier, Jr.
President, Moravian College

Dr. Morton Corn, Professor and Director
Division of Environmental Health Engineering
The Johns Hopkins University School of Hygiene and Public Health

Dr. John Doull
Professor of Pharmacology and Toxicology
University of Kansas Medical Center

Dr. Jack D. Hackney
Chief, Environmental Health Laboratories
Rancho Los Amigos Hospital Campus of the University of Southern California

Dr. Marvin Kuschner
Dean, School of Medicine
State University of New York at Stony Brook

Dr. Daniel Menzel
Director, Cancer Toxicology and Chemical Carcinogenesis Program
Duke University Medical Center

Dr. D. Warner North
Decision Focus, Inc., Los Altos, California

Dr. William J. Schull
Director and Professor of Population Genetics
School of Public Health
University of Texas at Houston

Dr. Michael J. Symons
Biostatistics Department

School of Public Health
University of North Carolina

Consultants

Dr. Seymour Abrahamson
Professor of Zoology and Genetics
Department of Zoology
University of Wisconsin at Madison

Dr. Edward F. Ferrand
Assistant Commissioner for Science and Technology
New York City Department of Environmental Protection

Dr. Ronald D. Hood
Professor, Developmental Biology Section
Department of Biology
The University of Alabama

Dr. Bernard Weiss
Professor, Division of Toxicology
School of Medicine
University of Rochester, New York

Special Consultants

Dr. Jack E. Gray
Department of Pathological and Experimental Research
The Upjohn Company

Dr. Samuel S. Lestz
Director, Center for Air and Biomedical Studies
Pennsylvania State University

Dr. Ronald Wyzga
Electric Power Research Institute
Palo Alto, California

NOTES

1. U.S. Environmental Protection Agency, "Control of Air Pollution from New Motor Vehicles and New Motor Vehicle Engines: Refueling Emission Regulations and Gasoline-fueled Light-Duty Vehicles and Trucks and Heavy-Duty Vehicles," *Federal Register*, vol. 52, no.160, August 19, 1987, p. 31165.

2. Ibid.

3. Ibid.

4. Ibid.

5. Section 202(a)(b) of the 1977 Clean Air Act Amendments.

6. *Federal Register*, p. 31163.

7. Ibid.

8. Ibid.

9. H. N. MacFarland et al., "A Chronic Inhalation Study of Unleaded Gasoline Vapor," *Journal of the American College of Toxicology*, 3, 1984, pp. 231–248.

10. Interview with Robert Scala, May 3, 1988.

11. Ibid.

12. Ibid.

13. Ibid.

14. Ibid.

15. Ibid.

16. U.S. Environmental Protection Agency, *Estimation of the Public Health Risk from Exposure to Gasoline Vapor via the Gasoline Marketing System*, Washington, D.C., June 1984 (hereinafter *Estimation*).

17. U.S. EPA, *Evaluation of Air Pollution Regulatory Strategies for Gasoline Marketing Industry*, Washington, D.C., July 1984 (hereinafter *Evaluation*).

18. *Estimation*.

19. International Agency for Research on Cancer, *Evaluation of Carcinogenic Risks to Humans*, World Health Organization, 1987, Supplement 7, pp. 31–32.

20. U.S. EPA, "Guidelines for Carcinogen Risk Assessment," *Federal Register*, vol. 5l, no. 185, 1986, pp. 33992–34003.

21. *Estimation*.

22. *Evaluation*.

23. *Estimation*.

24. U.S. EPA Science Advisory Board, Environmental Health Committee, *Transcript*, July 25, 1984. All quotations in this section are from these transcripts.

25. "EPA Arm Recommends End to Gas Vapor/Cancer Inquiry," *Automotive News*, August 6, 1984, p. 1.

26. Interview with David Cleverly, April 3, 1987.

27. Interview with Peter Preuss, May 15, l987.

28. Interview with Terry Yosie, May 15, 1987.

29. Letter from Fred Bowditch, Vice President, MVMA, to Herschel Griffin, Chairman, Environmental Health Committee, October 17, 1984.

30. Letter from Herschel Griffin, Chairman, Environmental Health Committee, and Norton Nelson, Chairman, SAB Executive Committee, to William Ruckelshaus, Administrator, U.S. EPA, October 29, 1984.

31. Science Advisory Board, Environmental Health Committee, "Environmental Health Committee Key Findings and Conclusions on the Draft Staff Paper, 'Estimation of the Public Health Risk from Exposure to Gasoline Vapor via the Gasoline Marketing System (June, 1984),' " October 1984, p. 5.

32. Letter from William Ruckelshaus, Administrator, U.S. EPA, to Herschel Griffin, Chairman, Environmental Health Committee, January 4, 1985.

33. Interview with Robert Scala.

34. Ibid.

35. Ibid.

36. Ibid., interview with Lawrence Slimak, July 10, 1987.

37. Interview with Robert Scala.

38. Ibid.

39. Ibid.

40. Interview with Lawrence Slimak.

41. Ibid.

42. Letter from Lawrence Slimak, Executive Secretary, Motor Vehicle Manufacturers Association, to Thomas Grumbly, Deputy Executive Director, Health Effects Institute, July 19, 1984.

43. Letters from Fred Bowditch, Vice President, Motor Vehicle Manufacturers Association, to Joseph Cannon, Assistant Administrator for Air and Radiation, U.S. EPA, and Bernard Goldstein, Assistant Administrator for Research and Development, U.S. EPA, July 19, 1984.

44. Interview with Lawrence Slimak.

45. Ibid.

46. Ibid.

47. Interview with Richard Paul, July 19, 1987.

48. Interview with Thomas Grumbly, August 14, 1987.

49. Interview with Stan Blacker, August 25, 1987.

50. Ibid.

51. Memorandum to Thomas Grumbly, Deputy Executive Director, HEI, from Robert Kavet, Staff Scientist, HEI, "Report on EPA Meeting on Gasoline Vapor (July 25, 1984)," August 1, 1984.

52. Letter from Thomas Grumbly, Deputy Executive Director, HEI, to Interested Parties, July 12, 1985.

53. Letter from Thomas Grumbly, Deputy Executive Director, HEI, to Board of Directors, HEI, April 23, 1985; emphasis in original.

54. Ibid.

55. Health Effects Institute, *Gasoline Vapor Exposure and Human Cancer: Evaluation of Existing Scientific Information and Recommendations for Future Research*, Report of the Health Effects Institute's Health Review Committee, Cambridge, Massachusetts, September 1985.

56. Ibid., p. 7.

57. Ibid.

58. Ibid., p. 8.

59. Ibid., p. 37.

60. Ibid., p. 8.

61. Ibid.

62. Ibid., pp. 38–39.

63. Ibid., pp. 39–40.

64. Ibid. p. 41.

65. Ibid.

66. Ibid.

67. Ibid.

68. Ibid., p. 42.

69. Interview with David Cleverly.

70. Interview with Terry Yosie.

71. Interview with David Cleverly.

72. Interview with Roger McClellan, July 6, 1987.

73. Interview with Robert Kavet, April 22, 1987.

74. Interview with Elizabeth Anderson, May 15, 1987.

75. Interview with Michael Walsh, June 10, 1987.

76. Letter from Bernard Goldstein, Chairman, Department of Environmental and Community Medicine, Robert Wood Johnson (Rutgers) Medical School, to John Graham, Associate Professor, Harvard School of Public Health, March 2, 1988.

77. Interview with Gary Martin, June 19, 1987.

78. Interview with Robert Kavet.

79. Interview with Thomas Grumbly.

80. Letter from Bernard Goldstein to John Graham.

81. Interview with Lee Thomas, August 13, 1987.

82. Personal interview, October 17, 1988.

83. U.S. EPA, *Evaluation of the Carcinogenicity of Unleaded Gasoline*, Washington, D.C., April, 1987.

84. U.S. EPA, *Draft Regulatory Impact Analysis: Proposed Refueling Emission Regulations for Gasoline-fueled Motor Vehicles*, Washington, D.C., July 1987.

85. U.S. EPA, *Evaluation of the Carcinogenicity of Unleaded Gasoline*, pp. 1–5.

86. Ibid., pp. 5–16.

87. U.S. EPA, *Draft Regulatory Impact Analysis: Proposed Refueling Emission Regulations for Gasoline-Fueled Motor Vehicles*, pp. 2–66.

88. U.S. EPA, "Control of Air Pollution from New Motor Vehicles," p. 31164.

89. Ibid., p. 31169.

90. Ibid., p. 31164.

91. Ibid., p. 31169.

92. Ibid.

93. Ibid.

94. Ibid., p. 31170.

95. Ibid.

96. Ibid.

CHAPTER 6

PERCHLOROETHYLENE

Elizabeth Drye

Tetrachloroethylene, also known as perchloroethylene, is a solvent used in metal degreasing, in dry cleaning, and as an intermediate in chlorofluorocarbon production. The chemical is particularly important to the dry-cleaning industry because no substitute exists that has both the low flammability and low acute toxicity characteristic of perchloroethylene.[1] The chemical was introduced to the industry in the 1940s as a substitute for a more toxic agent, carbon tetrachloride. As of 1985 about 75 percent of U.S. dry cleaners used perchloroethylene, accounting for about 51 percent of the 322,000 tons of perchloroethylene consumed annually in the United States.[2]

Perchloroethylene has been considered for regulation or has been regulated by several federal agencies. The focus of recent activity, however, has been at the Environmental Protection Agency (EPA). Several EPA program offices, including those administering air, water, and hazardous waste legislation, have evaluated or are evaluating the carcinogenicity of the chemical and have proposed and/or passed regulations governing its use and disposal.

Central to the EPA's regulatory deliberations has been the agency's development of the *Health Assessment Document for Tetrachloroethylene*.[3] A health-assessment documemt (HAD) purports to review relevant research and draw conclusions about a chemical's acute and chronic effects on human health. Health-assessment documents are routinely developed as multimedia support documents intended for use across agency programs and serve both as a key input into regulatory decision making within the agency and as a resource for other agencies at the local, state, federal, and international levels.

The EPA's Science Advisory Board (SAB) has been involved extensively in the development of the HAD for perchloroethylene. SAB review of HADs expected to be used as the basis of regulation is routine,[4] and before the HAD for perchloroethylene was finalized in 1985, the SAB had reviewed drafts of part or all of the document on three separate occasions. In May 1986 the Halogenated Organics Subcommittee of the SAB's Environmental Health Committee reviewed a draft addendum to the document that was intended to incorporate research results released after the original HAD had been finalized.

A central issue throughout all of these reviews has been the proper characterization of the chemical's carcinogenic potential in humans. The agency's qualitative assessment of perchloroethylene's carcinogenic potential and, to a lesser extent, the agency's quantitative estimate of cancer risk for the chemical are widely perceived to be crucial in determining the extent to which perchloroethylene should be regulated by the agency. In the current debate over the proper classification of perchloroethylene under EPA's Cancer Risk Assessment Guidelines,[5] the choice between the two categories being considered—"possible human carcinogen" and "probable human carcinogen"—is anticipated to have major regulatory implications. For example, the category assignment is expected to influence, if not determine, whether perchloroethylene is listed and regulated by EPA as a hazardous air pollutant, and whether the maximum contaminant level goal (MCLG) for perchloroethylene in drinking water is set to zero. The classification is also expected to have impacts outside of EPA; a classification of "probable" is anticipated to increase liability for perchloroethylene producers and users and to trigger increased state regulation of the chemical.

As we shall see, a variety of opinions exist about what human cancer risks are implied by the health effects data available on perchloroethylene. At each stage in the document's development, the carcinogenicity debate has been shaped by both the evolving scientific evidence and the agency's guidelines for cancer risk assessment. EPA's approach to classifying chemical carcinogens has also evolved over the past ten years, and these changes have influenced the language used in the debate over perchloroethylene's classification as a carcinogen.

This chapter examines in the case of perchloroethylene the influence of scientific advisory mechanisms on the EPA's risk-assessment and management processes. The first part of the chapter describes how the SAB operated in this particular case, and how its participation in the development of the health-assessment document for perchloroethylene influenced the agency's assessment of perchloroethylene's carcinogenicity and related administrative actions. The second part of the chapter offers some observations about the SAB's work in this context.

HISTORY OF THE SAB'S WORK ON PERCHLOROETHYLENE

Initial Health Assessment and SAB Review

In the late 1970s the EPA's Office of Air and Radiation requested that the Environmental Criteria and Assessment Office (ECAO)—a unit in the agency's research bureaucracy—develop a perchloroethylene HAD. Because of the substance's widespread production and use and its perceived potential to cause adverse health effects, the substance ranked high on the Air Office's list of chemicals to be considered for listing and regulation as hazardous air pollutants.[6] The major health endpoint of concern was cancer, and at the request of the ECAO, the agency's Carcinogen Assessment Group (CAG) wrote a section on carcinogenicity for inclusion in the draft document.

To characterize perchloroethylene's human carcinogenic potential, CAG employed the agency's 1976 proposed Cancer Risk Assessment Guidelines.[7] Under its 1976 guidelines the agency had essentially a four-category classification system. Evidence of human carcinogenicity was either "best," "substantial," "suggestive," or in none of these three categories. According to the proposed guidelines, epidemiologic studies provide the "best" evidence of carcinogenicity. "Substantial" evidence "that the chemical is a human carcinogen . . . is provided by animal tests that demonstrate the induction of malignant tumors in one or more species including benign tumors that are generally recognized as early stages of malignancy." Suggestive evidence "includes the induction of only . . . benign tumors . . . and indirect tests of tumorigenic activity."[8] In its carcinogenicity section for the perchloroethylene HAD, CAG concluded that "substantial" evidence of human carcinogenicity existed.[9]

CAG's conclusion was based largely on the results of a single, long-term animal bioassay conducted by the National Cancer Institute.[10] In the NCI bioassay perchloroethylene was administered in corn oil by gavage (stomach tube) to mice and rats of both sexes at three dose levels. Perchloroethylene exposure was associated with a significant increase in liver tumors in both sexes of mice. The mice results are summarized in Table 6.1. Excessive mortality in the rat study precluded drawing conclusions from the rat results.

According to CAG's interpretation, some additional research reviewed in the document provided further evidence of carcinogenicity. Results of epidemiology studies were inconclusive—inadequacies in classifications of exposures and control for exposures to other solvents precluded drawing conclusions based on study results. CAG felt, however, that in spite of its exhibiting the typical limitations, an NCI epidemiology study published in a preliminary report provided "suggestive evidence" of carcinogenicity.[11] Research demonstrating that perchloroethylene is converted to an epoxide metabolite thought to have carcinogenic potential also lent support to CAG's conclusion.[12] Furthermore, from CAG's point of view, "negative" study results did not contradict the NCI bioassay data. Mutagenicity data was "limited and inconclusive,"[13] and the negative results of both the NCI rat bioassay and a second rat bioassay conducted by Dow Chemical Company had deficiencies that made them "insufficient to assess the lack of carcinogenic effect in rats."[14]

In September 1980 the cancer risk-assessment documents for perchloroethylene and for five other volatile organic compounds were reviewed by the SAB's Subcommittee on Airborne Carcinogens. The subcommittee felt that overall the quality of the documents was unacceptable and that specifically the documents

Table 6.1
Incidence of Hepatocellular Carcinomas in Mice in the NCI Gavage Study

MALES		FEMALES	
Time-weighted average gavage dose (mg/kg/day)[a]	Incidence	Time-weighted average gavage dose (mg/kg/day)	Incidence
Untreated	2/20	Untreated	2/20
O (vehicle)[b]	2/20	0 (vehicle)	0/20
536	32/49	386	19/48
1,072	27/48	772	19/48

Source: U.S. Environmental Protection Agency, *Health Assessment Document for Tetrachloro-ethylene*, Washington D.C., 1985, p. 9–4.

[a]Sum (dose in mg/kg × no. of days at that dose)
Sum (no. of days receiving any dose)

[b]This control group given vehicle (corn oil) used to administer perchloroethylene to treatment groups.

should include "more information on the scientific protocols used in the experiments and studies cited in the documents."[15] The subcommittee concluded that the data presented in the perchloroethylene HAD did not support CAG's conclusion of substantial evidence for human carcinogenicity.[16] The relevant question, the subcommittee claimed, was not how much evidence for carcinogenicity had been presented, but whether the document contained any such evidence at all.[17]

In its review of the perchloroethylene document, the subcommittee discussed several weaknesses in the perchloroethylene data. In particular, certain characteristics of the NCI study design brought into question CAG's interpretation of the study's results. The cause of the excess tumors was questioned because a technical grade of the chemical contaminated with a potentially carcinogenic stabilizer rather than a pure grade of the substance had been administered to test animals. Subcommittee Chairman Sydney Weinhouse and subcommittee member Kenneth Baker advanced the argument that the presence of the stabilizer precluded inferring from the data that perchloroethylene had induced the excess liver tumors.[18] The vehicle of administration was also viewed by Baker as a potential source of spurious results.[19]

The most contentious debate, however, concerned the relevance of the tumor type induced to human risk assessment. Interpretation of liver tumor induction in the B6C3F1 mouse was widely debated in the scientific community. Some scientists believed that induction of this tumor type could be attributed in certain instances to recurrent liver tissue damage rather than carcinogenesis; because the strain has a high background rate of tumors in liver tissue, chemicals causing liver damage leading to increased tissue growth in this species could increase tumor incidence without acting as direct carcinogens. To some subcommittee members this mechanism provided the most plausible explanation for the NCI perchloroethylene bioassay findings, particularly in light of the equivocal mutagenicity evidence for perchloroethylene and the absence of an observed relationship between the doses of perchloroethylene administered in the NCI study and the magnitude of the tumor responses. Since humans do not exhibit the high background liver tumor rate on which this explanation of the NCI mouse data depended, some scientists questioned the relevance of the NCI findings to human risk estimates. The agency, however, consistently chose to view the induction of mouse liver tumors as evidence of carcinogenicity, and CAG's interpretation of the perchloroethylene data was consistent with this approach.[20]

The subcommittee also criticized CAG's quantitative assessment of perchloroethylene's carcinogenicity. According to CAG scientist Charles Hiremath, Subcommittee Chairman Sydney Weinhouse "strongly objected" to the agency's calculating an estimate of the excess lifetime human cancer risk attributable to a lifetime of continuous exposure to one unit of perchloroethylene, or a "unit risk" estimate.[21] Given the absence of an observed dose-response relationship in the NCI data, the panel felt that the agency's application to the data of a linear, nonthreshold, one-hit model for estimating unit risk was inappropriate.

In presentations following the subcommittee's discussion of the perchloroethylene assessment, representatives from perchloroethylene manufacturers and dry-cleaning associations objected to both CAG's quantitative and qualitative carcinogenicity assessments. Industrial scientists echoed the arguments of those board members who believed that recurrent liver damage rather than carcinogenesis best explained the NCI bioassay results; further, they offered additional rationales for rejecting both the EPA's unit risk-estimate and the agency's classification of the evidence as "substantial."[22]

In its review of the six chemical risk assessments, the SAB subcommittee criticized not only the agency's conclusions, but also the use of the terms "substantial" and "suggestive" evidence to characterize the carcinogenic potential of the chemicals. The board's unwillingness to frame the evidence for perchloroethylene's carcinogenicity according to the agency's 1976 proposed guidelines reflected a perceived disparity between the strength of the evidence associated with the terms "suggestive" and "substantial" in the guidelines and the literal meaning of these terms.[23] But the board did not offer an alternative approach. The subcommittee denied responsibility for evaluating the guidelines because "that was in the realm of regulatory policy" and recommended that the guidelines "be reviewed by an independent scientific group such as the National Academy of Sciences."[24]

The SAB's rejection of the use of the "substantial" and "suggestive" labels was perceived by Roy Albert, acting director of CAG and principal author of the guidelines, to have direct regulatory implications. He informed the committee, "If you say a chemical is less than 'substantial' or 'suggestive,' it will stop a standard cold."[25] Indeed, in 1981 consideration of perchloroethylene regulations under Section 112 of the Clean Air Act was deferred. According to Mark Greenburg of ECAO, the deferral was in part attributable to the subcommittee's apparent rejection of the guidelines as well. Recently, he reflected: "The agency was not going to make a decision on [the six halogenated organic] chemicals until they had an established set of guidelines. Our case [for regulating perchloroethylene as a carcinogen] would be stronger with more data and with guidelines."[26] The deferral, however, may have been caused by the change in agency management as well as by the outcome of the SAB review. Between the time Administrator Anne Gorsuch of the Reagan administration took office in the spring of 1981 and her resignation two years later, no new substances were listed for regulation under Section 112.[27]

Redrafting of the HAD and the 1982 SAB Review

The SAB again reviewed CAG's cancer risk assessment for perchloroethylene when the HAD went before the board's Environmental Health Committee, first in 1982 and subsequently in 1984. Between the 1980 and 1982 reviews, CAG's assessment of perchloroethylene's carcinogenicity had been modified and incorporated into the document. Most significantly, the evidence for perchloroethylene's carcinogenicity had been reexpressed using the classification scheme of the International Agency for Research on Cancer (IARC).[28] Based mainly on its assessment of the evidence for carcinogenicity from animal and human studies, IARC categorized chemicals in one of three major groups—"human carcinogen" (group 1), "probable human carcinogen" (group 2), and "cannot be classified . . . as to its carcinogenicity in humans" (group 3). The combination of the strength of the animal evidence (inadequate, limited, or sufficient) and the strength of the human evidence (inadequate, limited, or sufficient) determined the group assignment (see Table 6.2). The draft concluded:

In summary, there is evidence that tetrachloroethylene is a potential human carcinogen, based on the definitive hepatocellular carcinoma response in both male and female mice and the metabolic evidence of a direct-acting epoxide metabolite. Applying the criteria for evaluation of experimental animal carcinogenesis studies used by the International Agency for Research on Cancer (IARC, 1979), this level of evidence would be regarded as "limited," meaning not sufficient to provide a firm conclusion on its carcinogenic potential in humans. Two studies of laundry and dry cleaning workers have shown

Table 6.2

Summary of IARC and EPA Categories for Classifying Evidence of Human Carcinogenicity

Strength of
Evidence

Human/Animal	Group	Symbol EPA	IARC
Sufficient	Human Carcinogen	A	1
Limited/Sufficient	Probable Human Carcinogen	B1	2A
Inadequate/Sufficient	Probable Human Carcinogen[a]	B2	2B
Inadequate/Limited	Possible Human Carcinogen (EPA) C Not Classifiable (IARC)		3
Inadequate/Inadequate	Not Classifiable	D	3
Less than two negative animal studies or one negative study each in animal and humans	Not Classifiable Evidence of non-carcinogenicity (EPA) Not Classifiable (IARC)	E	3

[a]In late 1987 IARC renamed group 2B "possible human carcinogen." See *Overall Evaluations of Carcinogenicity: An Updating of IARC Monographs Volumes 1 to 42*, Supplement 7, International Agency for Research on Cancer, Lyon, France, 1987.

preliminary evidence of excess cancer incidence associated with tetrachloroethylene exposure, but no conclusion can be drawn from these until they are complete.[29]

Clearly, CAG conveyed less confidence that perchloroethylene was a human carcinogen with this statement than it had in its conclusion to the 1980 draft. In contrast to the earlier review, the SAB panel did not object outright to the level of certainty CAG had reached about perchloroethylene's carcinogenic potential; rather, the board expressed concern about the ambiguity of CAG's conclusions. In an attempt to evaluate the document from the perspective of those who would be using it, SAB member Herschel Griffin concluded that the word "potential" was unclear, the word "limited" was misleading to those unfamiliar with the IARC guidelines, and the mention of "preliminary results" from incomplete epidemiology studies was inappropriate. He summarized: I suppose what I am really saying is that I cannot tell from [the HAD] at this time, from the words that are there, assuming now that I am a policy analyst or something, whether or not the agency believes this is a carcinogen or not."[30]

The board reserved its most directive comments for CAG's unit risk estimate.

The discussion on perchloroethylene followed a presentation by the agency on the EPA's adoption of the linear, multistage model as its standard method of deriving an upper-bound estimate of low-dose cancer risks from dose-response data. The model provided a "plausible upper-bound" estimate the agency felt to be useful in regulatory decision making. Although accepting the methodology in general, Warner North objected to the application of the model to the perchloroethylene data. He stated:

Some method of assessing relative potency for carcinogens is needed, and an upper-bound calculation using consistent assumptions may be useful, especially for establishing Agency priorities. However, the existing CAG methodology is not a panacea for dealing with highly complex biomedical issues. Assumptions have to be tailored to scientific information for the specific chemicals if quantitative assessments are to be credible for the basis for regulatory decisions. . . . The aspect that concerns me particularly is the use of conservative assumptions. How much conservatism is appropriate, is the central question for risk assessment, and it's a question that requires both scientific and policy judgments.[31]

In particular, North argued that the choice of a no-threshold model was inconsistent with the fact that the document did not provide any evidence that perchloroethylene has direct genetic effects.[32] Additionally, he suggested that the assumption that humans are as sensitive to perchloroethylene as mice may be too conservative since pharmacokinetic evidence presented in the document indicated that mice are "much more sensitive than humans" to perchloroethylene exposure.[33] North concluded that the use of these assumptions potentially overstated risk and suggested that the unit risk estimate be compared to epidemiological data to verify that the estimate was indeed "plausible."[34]

Ultimately, the committee suggested that if the agency chose to include the unit risk estimate, it should put the potency of the substance in context. "The committee agreed that the EPA should include in its health assessment documents . . . the number of people exposed to the substance; the potency of the substance; and a comparison of the unit risk of a substance to the unit risks of other substances, such as arsenic and dioxin."[35]

Uncomfortable with the use of the NCI study results as the primary basis for the document's carcinogenicity conclusions and aware that research was under way that might clarify perchloroethylene's carcinogenic properties, several members of the committee recommended that the document's publication be deferred. In particular, a National Toxicology Program inhalation bioassay, conducted because certain characteristics of the 1977 NCI study (e.g., administration by gavage) limited the scientific community's ability to make sound inferences from study results, was nearing completion. However, Elizabeth Anderson of CAG explained that in part because of pressure "being put on the Office of Air Programs to reach some decision on the information available," a better alternative would be to finalize the document, even "if the answer is we simply can't make a determination," and publish an addendum to the document when new information became available.[36] Ultimately, this was the approach taken by the agency.

The Second Environmental Health Committee Review

Accordingly, the agency again revised the draft and in 1983 circulated the new version of the HAD for public comment.[37] Comments on the previous draft by SAB and the public had to some extent been incorporated into the draft. Notably, an appendix highlighting the differences between methods for extrapolating cancer risk estimates from high to low doses had been added. In addition, the agency had more precisely explained its qualitative carcinogenicity conclusions.

In the updated document the agency characterized the animal evidence for perchloroethylene's carcinogenicity as "limited" and the epidemiological evidence as "inconclusive" for assessing human cancer risk.[38] The combination of limited animal evidence and inadequate epidemiological evidence placed the chemical in IARC group 3—"the chemical . . . cannot be classified as to its carcinogenicity in humans" (see Table 6.2). However, CAG modified its carcinogenicity conclusions shortly after the draft's release; on March 21, approximately two months prior to SAB review of the new draft, the agency published a *Federal Register* notice modifying this conclusion.[39] In the notice the agency changed its characterization of the animal evidence from limited to "nearly sufficient." According to IARC rules, a chemical with sufficient animal evidence and inadequate epidemiological evidence is placed in IARC group 2B—"the chemical is a probable human carcinogen." Hence the *Federal Register* notice read: "The chemical falls in IARC category 3, but is close to a 2b, i.e. the more conservative scientific view would regard tetrachloroethylene as being close to a probable human carcinogen."[40]

In May 1984 the Environmental Health Committee again reviewed the draft perchloroethylene HAD. In spite of the fact that the agency had to some degree incorporated SAB comments on the previous draft into the new document, some members of both the committee and the industry groups present at the review felt that the agency had been unresponsive to the board's earlier suggestions.[41] Committee member Warner North expressed frustration that several simple editing changes he had recommended earlier had not been incorporated. Industry groups felt that the agency had been particularly unresponsive to the board's earlier comments on the document's treatment of carcinogenicity issues. As a Halogenated Solvents Industry Alliance (HSIA) representative commenting on both the perchloroethylene and trichloroethylene documents stated:

While the documents are comprehensive and provide, for the most part, a good review of the data base, the carcinogenicity sections are where we have a problem. They don't reflect past reviews by this committee in 1982 or by the Subcommittee on Airborne Carcinogens in 1980. In each of those cases, CAG came and asserted that there was

substantial evidence or sufficient evidence of carcinogenicity. In each case those committees said that they thought the evidence as to carcinogenicity was so weak that it
couldn't support CAG's conclusions, and different members of those committees expressed that they had a great lack of confidence in the cancer risk estimates. Yet the
documents are back, for the third time now, using the same data to assert that [trichlo-
roethylene] and perchloroethylene are probable or close to probable human carcinogens,
and for the third time they're using the same data, the same 1977 NCI study to compute
. . . precise estimates of human cancer risk.[42]

The late release of the *Federal Register* notice was criticized by both industry and the SAB. The timing of the notice evoked suspicion from industry
and criticism from SAB members, some of whom had not received the notice
at the beginning of the meeting. William Fisher of the Dry Cleaning Council
pointed out that the change followed suspiciously on the "heels of a published report by EPA's drinking water office to push for regulation of perchloroethylene as a carcinogen."[43] Likewise, committee member Herman
Collier voiced his concern that "it [was] rather easy to get the perception that
maybe the agency is posturing itself to, in fact, produce an answer that it already thought necessary to have."[44]

The phrasing of the conclusions in the notice also evoked criticism from the
board. Several committee members were uncomfortable with the use of the word
"conservative" in the context of the category assignment. As Chairman Griffin
stated, "Some time, perhaps in another forum, we are going to discuss what a
conservative scientific view is. . . . some would say that this is not a conservative
scientific view, but a conservative political view."[45]

This concern and others discussed at the meeting were reflected in the board's
letter to the EPA administrator.[46] As they stated in the letter, the board members
felt that "the critical issue" addressed by the Environmental Health Committee
in its review of the HAD was that of the chemical's carcinogenicity. The board
characterized the mouse liver tumors as a "signal" for the need to be concerned
about adverse health effects but was unable to provide an "authoritative scientific
opinion" on the compound's carcinogenicity. All but one committee member
agreed that the animal evidence was insufficient and that the chemical belonged
in IARC category 3, possible human carcinogen; one member agreed with the
conclusions, as stated in the draft, that the compound almost belonged in IARC
2B, probable human carcinogen.

As in the previous reviews, the committee was uncomfortable with the presentation of a quantitative risk estimate. It concluded that the evidence for
carcinogenicity was too weak to support a "formal calculation of a unit risk
estimate on purely scientific grounds" and recommended that the agency present
the calculation explicitly as a "what if" calculation to emphasize that the calculation was only meaningful if one assumed that perchloroethylene actually is
a human carcinogen. The committee also recommended that the rationale for
the risk assessment be expanded, and that the document address the suitability

of the "ambiguous bioassay" data to EPA's methodology. Further, the committee recommended that a comparison of the unit cancer risk estimate and results of epidemiological studies be made to discern whether or not the results were compatible. Finally, conditional on these and other more minor changes, the committee felt that the document would be "scientifically adequate" as a source document for agency use.

Redrafting of the Cancer Risk-Assessment Guidelines

The agency's finalization of the perchloroethylene HAD overlapped in time with its rewriting of its 1976 proposed Cancer Risk Assessment Guidelines. Work on the guidelines revision had begun in January 1984, and the proposed guidelines were published in the *Federal Register* in November of that year.[47] SAB members participated in development of the guidelines, and the guidelines were reviewed by the SAB in March and April 1985. The final guidelines, incorporating SAB and public comments, were published in September 1986.[48]

Revision of the guidelines was undertaken to incorporate advances in the risk-assessment field that had occurred since they had first been proposed in 1976.[49] The new guidelines set forth the agency's approach for calculating numerical estimates of risk and for "categorizing the weight of evidence for carcinogenicity from human and animal studies."[50] The new weight-of-the-evidence classifications represented a dramatic change from those proposed in 1976; the agency abandoned its categories of "suggestive" and "substantial" and adapted its category system from the IARC scheme.

Under the new guidelines the agency's method of characterizing the evidence for carcinogenicity from animal and human studies closely corresponded to that of IARC, as did its method of deriving an overall carcinogenicity category for the chemical; however, the classification system differed notably from IARC's in two respects. First, in categorizing the chemical, the guidelines used a "weight of-the-evidence" approach, which required consideration of all relevant evidence in the final category determination. IARC, in contrast, did not consider the implications of negative epidemiological and animal studies or mechanistic data in making category assignments. Second, while retaining categories corresponding to IARC's group 1 and group 2, the agency substituted three more detailed categories for IARC group 3 to further distinguish among the broad range of chemicals appropriately assigned to that group. Consequently, in contrast to the IARC scheme, the agency not only has categories for human carcinogens (group A), probable human carcinogens (group B), and chemicals that cannot be classified (group D), but for "possible human carcinogens" (group C) and for chemicals for which "evidence of non-carcinogenicity" exists (group E) as well. A comparison of the IARC and agency categories is provided in Table 6.2.

Table 6.3
Combined Incidence of Hepatocellular Adenomas and Carcinomas in Mice in NTP Inhalation Study

Exposure Group	Incidence	
	Male	Female
Control	17/49	4/48
100 ppm	31/49	17/50
200 ppm	41/50	38/50

Source: U.S. Department of Health and Human Services, National Institutes of Health, National Toxicology Program, *NTP Technical Report on the Toxicology and Carcinogenesis Studies of Tetrachloroethylene in F344/N Rats and B6C3F1 Mice*, NTP TR 311, Research Triangle Park, North Carolina, August 1986, p. 53.

In the interim period between the publication of the proposed and final guidelines, the agency adopted informally the proposed guidelines for use in ongoing assessments. Hence the final HAD, released in July 1985, applied both the IARC and EPA carcinogenicity classifications to perchloroethylene. Based on the conclusion that limited animal evidence and inadequate human evidence of carcinogenicity existed, agency scientists assigned perchloroethylene to EPA weight-of-the-evidence group C and IARC group 3—assignments that also reflected the opinions expressed by Environmental Health Committee members at the 1984 review.

Likewise, the document's presentation of the quantitative risk estimate reflected SAB comments. The risk estimate was presented in the context of a "what if" scenario, and the rationale for the agency's choice of risk model was expanded. Further, a comparison was made between the unit cancer risk estimate and epidemiological data, indicating that the two were not incompatible. After three SAB reviews spanning five years, the agency had finally arrived at a characterization of perchloroethylene's human carcinogenic potential that was acceptable to the agency's Science Advisory Board.

NTP Study Results

Shortly after the approved HAD was published, the long-anticipated results of the National Toxicology Program (NTP) animal bioassay were released.[51] In the NTP study male and female B6C3F1 mice and male and female F344 rats were administered lifetime exposures to perchloroethylene by inhalation. The tumor incidences for mice and rats are summarized in Tables 6.3 and 6.4,

Table 6.4
Incidence of Mononuclear-Cell Leukemias and Renal Tumors in Rats in NTP Study

Exposure Group	Mononuclear-cell leukemia	Renal Tumors Tubular-cell adenomas	Tubular-cell carcinomas	Combined Incidence
Males				
Control	28/50	1/49	0/49	1/49
200 ppm	37/50	3/49	0/49	3/49
400 ppm	37/50	2/50	2/50	4/50
Females				
Control	18/50	0/50	0/50	0/50
200 ppm	30/50	0/49	0/49	0/49
400 ppm	29/50	0/50	0/50	0/50

Source: U.S. Department of Health and Human Services, National Institutes of Health, National Toxicology Program, NTP Technical Report on the Toxicology and Carcinogenesis Studies of Tetrachloroethylene in F344/N Rats and B6C3 F1 Mice, NTP TR 311, Research Triangle Park, North Carolina, August 1986, pp. 41–43.

respectively. As in the 1977 NCI bioassay, perchloroethylene exposure was associated with an increased incidence of hepatic tumors in both male and female mice. A statistically significant increase of mononuclear-cell leukemia in exposed rats was observed for both sexes at both doses. Male rats in the exposed groups also exhibited a dose-related increase in incidence of renal adenomas and carcinomas.

The strength of the mouse response, as evidenced by the magnitude of the increase in hepatocellular tumor incidence and the clear dose-response relationship, verified the results of the earlier NCI bioassay and reinforced the conclusion that perchloroethylene exposure induces liver tumors in both sexes of the mouse. The rat results were not as readily interpretable. Because of the type and number of tumors involved, interpretation of the rat results was widely debated within the scientific community. The increase in kidney tumors in male rats was not statistically significant; however, the rare nature of the tumor type and the dose-related trend suggested the existence of an important effect. In

contrast, the relatively high background incidence of mononuclear-cell leukemia in rats cautioned against attributing an elevation in the incidence of these
tumors among exposed groups to perchloroethylene exposure. Further, in the
view of some scientists, the pathology for this tumor type was not well enough
established to allow sound inferences from the incidence rates identified in the
NTP study.

In August 1985 the National Toxicology Board of Scientific Counselors, a
peer review board of the National Toxicology Program, weighed these considerations and concluded that the study provided "clear evidence of carcinogenicity" in both sexes of mice and in male rats, and "some evidence of
carcinogenicity" in female rats. The decision to place the male rat evidence in
the "clear" category, however, was passed by a narrow margin (five votes in
favor, four votes against).

Disagreement on the male rat evidence reflected differences in interpretation
of the mononuclear-cell leukemias. "Clear evidence," by definition, is provided
"by studies that are interpreted as showing a chemically related increased incidence of malignant neoplasms [tumors], studies that exhibit a substantially
increased evidence of benign neoplasms, or studies that exhibit an increased
incidence of a combination of malignant and benign neoplasms where each
increases with dose."[52] One panelist voting against a "clear" classification
refuted the importance of the leukemias by pointing to the high background
incidence observed in control animals. Another panelist favored the "clear"
categorization given that the increase in rat leukemias was statistically significant.[53]

Agency Response

The release of the NTP study results and the Board of Scientific Counselors'
conclusions prompted Elizabeth Anderson, then director of the Office of Health
and Environmental Assessment (OHEA), CAG's parent office, to distribute an
internal memo on September 29, 1985.[54] The memo indicated, "in an interim
manner," the study's findings and its effect on the chemical's weight-of-the-
evidence classification and announced the agency's intent to draft an addendum
to the perchloroethylene HAD incorporating the NTP results. The memo concluded:

In consideration of both the new NTP rat and mouse inhalation bioassay showing positive
carcinogenic evidence in both species and the positive carcinogenic evidence in mice by
gavage, the overall evidence for carcinogenicity in animals is now elevated to the "sufficient" category. Using the Agency's proposed guidelines for carcinogen risk assessment,
the overall carcinogenic evidence for tetrachloroethylene would be elevated to Group B2
meaning that tetrachloroethylene should be considered a "probable" human carcinogen.[55]

Following the circulation of the Anderson memo and prior to release of the
draft addendum for notice and comment, two agency program offices acted on

the new information. In light of the study results, the Office of Drinking Water reopened the comment period on the proposed recommended minimum contaminant level (RMCL) for perchloroethylene,[56] and the Air Office published in the *Federal Register* a notice of intent to list perchloroethylene as a hazardous air pollutant.[57] The Air Office's notice cited the Anderson memo, stating that there was a "high likelihood that [perchloroethylene] will be a probable human carcinogen under the new EPA guidelines."[58]

The Addendum and the Halogenated Organics Subcommittee Review

In April 1986 the draft addendum to the perchloroethylene HAD was released for public comment.[59] The document summarized the NTP inhalation study and modified the agency's carcinogenicity findings based on the results. The document concluded that perchloroethylene was a B2 or "probable" human carcinogen under EPA's proposed guidelines and provided three ranges of unit risk estimates for inhalation exposure, each calculated using three different methods. The first, calculated using the NTP study's mouse liver tumor and rat leukemia responses and EPA's standard methodology, gave a result comparable to that presented in the 1985 HAD (2.9×10^{-7} to 9.5×10^{-7} per $\mu g/m^3$ exposure). The second and third ranges, calculated from the same tumor results but using a method that incorporated pharmacokinetic information, gave somewhat higher risk estimates (2.9×10^{-6} to 1.1×10^{-5} per $\mu g/m^3$ exposure; 9.6×10^{-7} to 3.6×10^{-6} per $\mu g/m^3$ exposure).

Despite the new NTP results, industrial commentators argued against classifying the animal evidence for perchloroethylene as "sufficient" (and hence against categorizing perchloroethylene as a B2 carcinogen). The following comments submitted by HSIA on the notice of intent to list are representative of those received by the agency from industry:

- The mouse liver tumors are not indicative of human cancer risk. Perchloroethylene induces tumors in mice through a nongenotoxic mechanism known as peroxisome proliferation, and the mechanism is not thought to exist in humans.[60]

- The results of the NTP rat bioassay are in doubt. "An audit of the study performed by Clement Associates for HSIA revealed startling deficiencies" in animal handling, indicated "failure to detect lesions in control animals," and suggested that the "staging criteria used [to evaluate the rat leukemias] may have overstated the incidence of leukemia."[61]

- The rat tumors may have little relevance to human risk assessment, since the leukemia is of a type not found in humans and the male rat is "peculiarly susceptible" to renal toxins due to unique characteristics of rat kidney physiology.[62]

On May 15, 1986, the draft addendum was reviewed by the SAB's Halogenated Organics Subcommittee, a subcommittee of the Environmental Health Committee. Since comments submitted to the agency had challenged the quality of the NTP study, the board spent some time in dialogue with John Minnear, an NTP scientist. Further, since the subcommittee did not have the draft NTP report to work with at the review, but only the EPA's presentation of the NTP's findings, some time was spent clarifying what findings had and had not been represented correctly in the draft addendum. Less than an hour of the meeting was spent in general discussion. As one agency scientist put it, "Most of the meeting was spent NTP-bashing."[63]

According to Ernest McConnell, director of the NTP's Division of Toxicology, Research, and Training, concerns about the study, prompted primarily by the findings of the Clement Associates audit and industry's emphasis on these findings, were largely unwarranted. McConnell felt that "any legitimate scientist would say that [the perchloroethylene bioassay] was a well conducted study" and that the study's level of quality was "in the top five percent." McConnell asserted that "deficiencies" in animal handling referred to in the report did not meaningfully affect exposures. Although somewhat understanding of industry's concern about the detection of the leukemias because of the qualitative nature of the pathology involved, McConnell insisted that industry's criticism of the staging criteria used to evaluate the tumors was pointless, as the tumors were significant regardless of whether or not the staging approach had been used.[64] Ultimately the subcommittee reached similar conclusions, and these issues were for the most part dismissed by SAB members.

During the brief period of general discussion in the late afternoon, the committee did not reach a clear consensus on its interpretation of the rat tumors or on the chemical's carcinogenicity classification. In the waning minutes of the meeting, the chairman of the subcommittee, John Doull, acknowledged that regardless of the significance of the rat tumors, the induction of mouse liver tumors by two routes of administration should place perchloroethylene in the weight-of-the-evidence category B2. Curtis Klaassen summed up the sentiment of the panel regarding the chemical's carcinogenicity:

> We were kind of caught in this dilemma: by the strict interpretation of the [guidelines] it should be considered a B2, and with a B2, by definition, that's probable. But we were saying with our scientific minds, it's probably not!
> I think that a lot of us left the room kind of frustrated in that regard. Here, probably EPA is going to call this a B2 because it fits this but we don't think there is a real human health problem there.[65]

Most agency scientists and industry representatives left the May meeting anticipating that the board would concur with the agency's classification of perchloroethylene as a probable human carcinogen. Indeed, troubled by this conclusion, Rudolf Jaeger, a consultant to Unifirst Corporation who had made a presentation to the board at the meeting, submitted a letter on May 19 to Doull expanding his views on the classification issue. The letter emphasized the flexibility of the guidelines, pointed out that "even where animal evidence is 'sufficient', the 'group 2B', 'probable human carcinogen' label is not *mandated*," and argued that the guidelines had been misapplied.[66]

Eight months after the May review, in a letter to Administrator Lee Thomas,[67] the board communicated its findings on the draft addendum to the agency. In the letter the board rejected industry's view that the way in which the NTP study had been conducted raised doubt about the rat results, yet it disagreed with the agency's interpretation of the rat findings. The board argued that aggregating adenomas and carcinomas in the analysis of rat kidney tumors was "not an obvious biological procedure" and pointed out that the trend for carcinomas alone was not dose-related. It noted as well that the statistical analysis of these tumors was in error. Likewise, the board placed less significance on the leukemia tumors than the agency, stating that "the scientific community has a poor understanding of [their] pathology" and referring to the agency's use of a three-stage classification scheme in its evaluation of the tumors as "preliminary and ad hoc." With respect to the mouse results, the board concurred with the agency that perchloroethylene induced a statistically significant increase in liver carcinomas. The letter emphasized, however, that the mouse results simply confirmed the findings of the earlier NCI study and concluded that "no new dispositive information has been gained." Finding no evidence of carcinogenicity in the rat results, and no basis for new inferences about the chemical's human carcinogenicity in the mouse results, the board disagreed with the position of agency scientists that the weight-of-the-evidence category should be B2, probable human carcinogen, and assigned it to group C, possible human carcinogen.

The board's challenge to some of the agency's scientific conclusions and its ultimate assignment of perchloroethylene to group C reflected the dilemma created by having to evaluate perchloroethylene using the framework set forth in the guidelines. As Doull described: "We were asked to look at the science, but we had a 'semantic problem.' . . . The subcommittee clearly felt that the category should be 'possible' and had to downplay the evidence to that effect."[68] Alternatively, the SAB could have classified the animal evidence as "sufficient" but recommended that the agency invoke the flexibility of the guidelines and assign the chemical to group C anyway. Instead, the SAB chose to characterize the scientific evidence in such a way that a straightforward application of the guidelines would lead to what was, in the panel's best judgment, the appropriate classification.

The letter also commented on the quantitative risk estimate. The board had long been pushing the agency to include in its HADs risk estimates calculated using a method that incorporated chemical-specific pharmacokinetic information. The perchloroethylene addendum represented the agency's first attempt to present this type of estimate, and the board applauded the agency for its efforts. As in its reviews of the 1985 HAD, the board also suggested that the quantitative risk estimate and results of epidemiological studies be evaluated for consistency.

Daniel Byrd, who, as executive secretary of the SAB, was primarily responsible for drafting the letter, summarized its tone and content:

The perchloroethylene report put [the board] into three different areas in three different subjects. It put them on the data interpretation question as being different than the agency, but sort of feeling on firm grounds and saying to the agency, "this is our interpretation, and we feel fairly confident. But if you guys want to differ with us that's o.k. . . . here's how you might go about constructing a rationale. On the quality of the document they were flame throwers, and on the [incorporation of pharmacokinetics in the risk estimate] they were complimenting the agency.[69]

From the points of view of some observers who had attended the public meeting, the board's letter to the administrator did not seem to accurately reflect the discussion at the subcommittee's May 1986 meeting. As one agency scientist who had participated at the meeting expressed, "The letter was a complete surprise." It was not clear to agency scientists how the SAB had arrived at some of its conclusions, and their ability to understand how the SAB had arrived at its position was limited somewhat by the fact that no transcript of the meeting had been taken.

OHEA scientists worked closely with NTP to address some of the issues raised both in the May Wisconsin meeting and in the letter and ultimately chose to formulate a detailed response to the board defending the carcinogenicity conclusions in the addendum. In a response to the SAB the agency defended its position on the significance of both types of rat tumors.[70] The agency asserted that combining rat kidney adenomas and carcinomas was both appropriate and consistent with the guidelines in this instance, since progression of the tumors from a benign to a malignant state "[had] not been discounted and, in this case, [was] likely." Refuting the board's position on the leukemia tumors, the agency expressed its concurrence with NTP scientists that "rat mononuclear cell leukemia is a well characterized disease," highlighted the insensitivity of the statistical results to the staging process, and cited the NTP's findings of "clear evidence" of carcinogenicity in support of its interpretation. Finally, with respect to the mouse results, the agency asserted that "the fact that a strong carcinogenic response [had] been demonstrated in two separate experiments, in different laboratories, using different routes of exposure, producing similar dose-related responses" added to the weight of the evidence for carcinogenicity in animals. Reaffirming its assignment of perchloroethylene to the weight-of-the-evidence category B2, the agency argued that the rat data, when considered in toto with the mouse data, or the mouse data alone supported this conclusion.

To bolster its overall conclusions, the agency pointed out that IARC was anticipated to elevate its classification of the animal evidence for perchloroethylene from limited to sufficient, based largely on the new evidence provided by the NTP mouse results. While IARC's position seemed to support the agency's conclusions, comparison of the agency's and IARC's positions was somewhat problematic. At the time, IARC was in the process of modifying slightly its approach to classifying carcinogens. Notably, the organization was changing the literal label associated with group 2B (the group roughly corresponding to EPA's B2) from "probable human carcinogen" to "possible human carcinogen." Group 2A continued to be associated with the description "probable human carcinogen."[71] The change meant that chemicals judged to have "sufficient animal evidence" and "inadequate human evidence" of carcinogenicity under both classification schemes would be labeled "possible human carcinogens" under IARC rules but "probable human carcinogens" under EPA guidelines. In light of this inconsistency, the agency recommended in its letter to the SAB that "the implications of [the] use of the terms 'possible' and 'probable' in EPA's risk assessments be a point for discussion among agency scientists and the SAB."

Although the SAB's placing of perchloroethylene in category C seemingly did not affect OHEA's conclusion regarding the weight of-the-evidence category, it did affect the activity of at least one of the agency's program offices. According to Penny Fenner-Crisp, director of the Office of Toxic Substances and Pesticides, "The [Halogenated Organics Subcommittee's] recommendation slowed down the rate at which [the office] worked on perchloroethylene."[72] In contrast, according to Jeanette Wiltse of the Air Office, consideration of regulation of perchloroethylene under the Clean Air Act "didn't stop moving" when the SAB subcommittee labeled the chemical a C.[73]

Public Response and Lee Thomas's Involvement

The agency's rejection of the Halogenated Organics Subcommittee's categorization for perchloroethylene resulted in direct pressure on Administrator Lee Thomas. Over two thousand letters from dry cleaners to their congressional representatives requesting that Congress encourage the agency to follow the SAB's advice were forwarded to the administrator, accompanied by letters from the congressmen themselves. Additionally, industry representatives arranged a meeting with the administrator to express their concern about the implications of the B2 classification.

In a June 1, 1987, letter to the administrator,[74] the HSIA listed the anticipated consequences both within and outside of the agency of the agency's labeling perchloroethylene a B2. The action, HSIA predicted, was likely to affect

- the maximum contaminant level goal (MCLG) set by the agency for the chemical under the Safe Drinking Water Act;

- whether or not the chemical was regulated under Section 112 of the Clean Air Act;

- whether or not the chemical was listed under the OSHA Hazard Communication Standard (29 C.F.R. 1910.1200);

- the level of the reportable quantity (RQ) for perchloroethylene established under Superfund;
- state regulations of perchloroethylene in drinking water and air; and
- the liability of perchloroethylene users and manufacturers.

The letter concluded:

Decisions based on the classification (e.g., to list under Section 112 of the Clean Air Act or to publish an MCLG of zero) have major adverse implications for industry and ultimately for the public. Such decisions create a precedent supporting the view that vanishingly small quantities of substances that are ubiquitous in the environment pose a danger to human health. . . . Hence, litigation is inevitable where regulatory decisions do not accurately reflect the weight of the evidence for carcinogenicity, even though the affected industry may have no particular quarrel with reasonable regulations to control emissions or discharges.

In response to this intense level of public comment, Thomas wrote a letter to the SAB on August 3, 1987, soliciting further advice from the board and expressing his concerns.[75] The letter solicited the board's opinion on the following three issues:

- The relative significance of rat kidney and mouse hepatocellular tumors for human risk assessment.
- The approach taken by EPA in using its guidelines to infer human carcinogenic potential from the total body of scientific evidence on perchloroethylene.
- Whether or not there was "research underway or anticipated that will clarify these rodent tumor responses and their relationship to human health risk assessment," and what additional research should be undertaken.

In his letter Thomas also expressed concern about the overriding importance given to the carcinogenicity classification by regulators. He pointed out that in some cases a group C substance might present a greater threat to human health than a group B substance, given that the difference in category assignments between the two substances might simply reflect a difference in the amount or quality of data available for each, and that the assignment did not capture the estimated potency of a substance or the number of people exposed to the substance.

Workshop on Mouse Liver and Rat Kidney Tumors, and SAB Response

On August 12, 1987, the SAB held a joint meeting of the Environmental Health Committee and its Halogenated Organics Subcommittee to review the current state of knowledge on mouse liver tumors and rat kidney tumors. Since the implications of these two tumor types for human risk assessment were widely debated yet critical to the evaluation of several organic compounds (including perchloroethylene), the board held the meeting to further its understanding of the extent to which these tumors were predictive of human risk. At the meeting recent research on possible mechanisms of carcinogenesis acting in the mouse liver and the rat kidney was reviewed, and the implications of the hypothesized mechanisms for human risk assessment were discussed. Based in part on discussions of the new research presented at the meeting, the board drafted a response to Lee Thomas's August 3 letter.

The response, which was not finalized until March 9, 1988, concluded that in the case of perchloroethylene the mouse tumors were better predictors of human risk than the rat tumors.[76] The board stated that the carcinogenic effect of perchloroethylene exposure on rat kidneys was likely to be attributable to the activity of a particular protein present in male rats but not in humans. Hence the rat kidney tumors "may not be relevant to human risk assessment."[77] In contrast, the board observed that the mechanism of carcinogenesis in mouse liver tumors was still not well understood and concluded that "the generation of mouse liver tumors by chemicals is an important predictor of potential risks to humans."

These conclusions seemed to be compatible with the agency's position that the chemical could be considered a probable human carcinogen based on the results of mouse liver tumors alone—a position that was reflected in the accompanying restatement of the board's opinion on perchloroethylene's classification under the EPA's Cancer Risk Assessment Guidelines. In response to Lee Thomas's request for the board's opinion on the agency's application of the guidelines to the chemical, the board stated:

The issues regarding the application of the risk assessment guidelines appear not to represent disagreement among scientists about scientific evidence but, rather, the consequence of attempting to fit the weights of evidence into necessarily arbitrary categories of risk. Since the weights of evidence, and uncertainties associated with such evidence, for perchloroethylene and other compounds fall within a range of scientifically defensible choices, it may not be possible, in some instances, to fit them neatly into only one risk category. Moreover, the more incomplete the data, the less precision one can expect in classifying a compound within EPA's cancer guidelines. In addition, the type of evidence that places a compound in a particular category may vary considerably from substance to substance within that category. For perchloroethylene, as with trichloroethylene, the Science Advisory Board concludes that the overall weight of evidence lies somewhere on the continuum between the categories B2 and C of EPA's risk assessment guidelines for cancer.[78]

The intent of the letter was in part to commend OHEA for its increased willingness to consider mechanistic data, such as the rat kidney data for perchloroethylene, in its assessment of human carcinogenic risk and to encourage the agency to be flexible in applying the guidelines. As Doull explained:

The SAB was saying to the agency, "Prudence is fine when you don't have the answer, but when you have the science then use it." . . . The Agency is always more conservative [than the SAB], but that's O.K. . . . If you read the guidelines, there is a lot of discussion about flexibility. But in the past this has been window dressing. We told the agency, "Figure out how to do what you're already committed to do."[79]

In the letter the board also expressed that it shared Lee Thomas's concern about the "black-and-white" regulatory implications of the B2 versus C distinction and noted that factors other than the "weight of the evidence" might suggest the need to regulate some C chemicals or might provide a "scientific justification" not to regulate some B2 chemicals.[80] Finally, the SAB recommended that the agency "reevaluate whether to change its labeling system and methods of characterizing uncertainty" both to improve the quality of the agency guidelines and to maintain consistency with IARC.[81]

Current Expectations

The agency has not yet reacted definitively to the SAB's March 9, 1988, letter. OHEA has not yet finalized the draft addendum. According to Penny Fenner-Crisp, the Office of Toxic Substances and Pesticides is waiting on the CAG's final carcinogenicity classification of perchloroethylene before moving ahead with further consideration of perchloroethylene regulation. She predicted that if the CAG recommends a B2, her office "will probably go with that." Fenner-Crisp pointed out, however, that the distinction between B2 and C is not the sole factor in determining whether or not her office decides to regulate, although that distinction "will probably affect the *strength* of the regulation(s)" promulgated.[82] Likewise, it does not appear that the board's most recent conclusions will necessarily affect the Air Office's decision on whether to regulate the chemical as a hazardous air pollutant. According to Jeanette Wiltse, "The office has been working under the assumption all along that the chemical was somewhere between a B2 and a C."[83]

In contrast, the agency's carcinogenicity classification will apparently determine the perchloroethylene standards promulgated by the Office of Drinking Water. The office regulates "possible" carcinogens less stringently than "probable" carcinogens; RMCLs for the latter are set at zero.[84] Likewise, the maximum contaminant level—the enforceable level—will be lower if perchloroethylene is placed in category B2 rather than in C. But the regulatory decision does not appear to be sensitive to the distinction between an agency conclusion of B2 and an agency determination that the chemical falls somewhere between the B2 and C categories. Given the latter conclusion, the Office of Drinking Water plans to err on the side of safety and regulate the chemical as if it were a probable human carcinogen.[85]

The comments of Fenner-Crisp and Wiltse counter some of industry's expressed concerns about the regulatory implications of the agency's final category determination. Unless CAG labels perchloroethylene a B2 carcinogen in its final document, it may never be clear whether industry's concerns about the response of these program offices and other regulatory agencies to the B2 categorization are well founded. However, it is generally anticipated that CAG will ultimately recommend a classification of B2, and that the recommendation will have important consequences both within and outside of the agency.

CONCLUSIONS

SAB review affected both the content and editorial quality of EPA's health-assessment document on perchloroethylene. SAB review improved the editorial quality of the HAD and is likely to positively affect the editorial quality of the addendum as well. This influence is evidenced by the multiple instances in which the board identified errors in the document and brought these to the attention of agency staff. The board also affected the document's editorial quality by pointing out text that was unclear and by encouraging the agency to be explicit both about its reasons for selecting a particular approach to estimating risk and about the assumptions made in the approach.

With respect to the agency's treatment of perchloroethylene's carcinogenicity in the 1985 HAD, the SAB exerted a restraining force on the agency. The board's influence on both the agency's conclusions about perchloroethylene's carcinogenicity and the way in which these conclusions were presented resulted in the agency's presenting a less conclusive statement about perchloroethylene's human carcinogenic potential in the HAD than it would have presented had the SAB not reviewed the document. The SAB's comments on the HAD prompted the agency to abandon its assessments of the evidence for perchloroethylene's carcinogenicity as "substantial" (under EPA's 1976 proposed guidelines) and "3 but close to a 2B" (under IARC rules) and to ultimately place the chemical in IARC group 3—a category denoting relatively less certainty about perchloroethylene's carcinogenic potential. The SAB's insistence that the agency present the quantitative risk estimate in the proper context (i.e., in the context of a "what if" scenario) further mitigated the strength of the agency's statements about the carcinogenic risk posed by perchloroethylene.

SAB review also affected the agency's regulation of perchloroethylene. A reasonable inference in this case is that perchloroethylene would have been subject to regulation earlier on had the SAB not been involved in its evaluation. To the extent that completion of the HAD was requisite to the agency's pursuing particular regulatory actions, the SAB's unwillingness to "approve" the document in its early years of development may have affected agency policy. Further, since SAB involvement caused the agency to make less conclusive statements about the chemical's carcinogenic potential in the 1985 HAD than it otherwise would have, SAB involvement served to discourage regulatory action during the first half of the 1980s.

The effect of more recent SAB involvement on regulatory decision making remains unclear. Since the agency has not finalized the draft addendum, it is unclear whether the board's deliberations will have any effect on the EPA's scientific conclusions and ultimately on the agency's regulatory policy. However, it appears that regardless of the agency's final weight-of-the-evidence category assignment for perchloroethylene, the SAB's categorization of the chemical first in category C and more recently as "somewhere on a continuum between the categories B2 and C" has slowed the pace at which regulation has been developed and promulgated.

The ultimate influence of SAB review on the agency's risk management of perchloroethylene will in part be a function of the sensitivity of the agency's regulatory policies to incremental improvements in its scientific information. It appears that if the agency ultimately recommends a B2 classification for per-chloroethylene, the SAB's comments on this issue will be weighed by the Air Office and the Office of Toxic Substances in their consideration of perchloro-ethylene regulation. In contrast, where the distinction between B2 and C is anticipated to drive regulation (e.g., in the Office of Drinking Water and among state agencies using the document), an agency determination of B2 will render the board's advice inconsequential.

The perchloroethylene case illustrates that the influence of the Science Advisory Board's decisions on risk management extends beyond the boundaries of the Environmental Protection Agency. As communicated to the administrator by the HSIA and as recognized by the administrator and the board, EPA's ultimate decision as to whether perchloroethylene is a "possible" or a "probable" human carcinogen is expected to influence not only EPA's regulatory policy, but the way in which the chemical is regulated by other federal and state agencies and, more fundamentally, how perchloroethylene is viewed by the general public. If the EPA's ultimate categorization of the chemical differs from what it would have been had the SAB not been involved in evaluating the perchloroethylene addendum, the SAB's influence will extend well beyond the boundaries of the EPA.

SAB review provided expanded opportunity for public involvement. It seems likely that in the absence of the SAB, industry would have been less involved in the agency's evaluation of perchloroethylene's health effects. SAB review provided a forum for public involvement, and industry apparently perceived heavy involvement at this stage of perchloroethylene's evaluation to be a strategic use of its resources. In contrast, the involvement of environmentalists in the agency's assessment of perchloroethylene's health effects has been minimal. This asymmetry is not surprising, given the relative areas of expertise of industry and environmental groups.

The implications of public involvement in the agency's development of the HAD and its addendum and of the asymmetry in the interests represented at this phase in perchloroethylene's evaluation are not clear. Public participation at this stage may have improved the quality of the agency's scientific determinations.

However, public presentations at SAB meetings consumed a significant amount of the board's time. Further, the asymmetry of the interests represented at SAB meetings may have biased the agency's and/or the SAB's conclusions in a direction favorable to industry interests. It is not obvious, however, that involvement at SAB reviews will ultimately increase industry's overall influence on the agency's management of the chemical as a human health risk over what it would have been had industry not been given the opportunity to participate in SAB reviews. For example, by participating in the agency's development of its scientific conclusion, industry may have diminished the possibility of its successfully challenging in the courts any agency decisions based on these conclusions.

The SAB's role in this case was not purely advisory. The ability of the SAB to function as an "advisory" institution in the perchloroethylene case was constrained by political realities. Throughout the perchloroethylene case agency scientists, regulators, and interest groups behaved as if SAB approval was requisite to the agency's use of the document's conclusions as an input into regulatory decision making. Hence, as Terry Yosie has noted,[86] in the perchloroethylene case the recommendations of the Halogenated Organics Subcommittee on perchloroethylene's classification in the addendum appear to have constrained the agency from exercising its latitude to form conclusions independent of the SAB's advice. To the degree that the SAB's advice narrowed the range of acceptable alternatives available to the agency, the SAB's "advice" lost some of its advisory character and may have in reality constricted the agency's actions. Moreover, given that the agency's assessment of perchloroethylene's human health effects is used by agencies outside the EPA, the influence of this constraining force potentially extended beyond the agency's walls.

The perchloroethylene case suggests that the SAB's role in science-policy determinations is expanding. It demonstrates an evolution in the type of advice the board believes that it should provide to the agency. Indeed, the evolution is in part a consequence of the board's having to evaluate perchloroethylene's carcinogenicity. The Subcommittee on Airborne Carcinogens' refusal to assume a role in evaluating the agency's Cancer Risk Assessment Guidelines reflected a sensitivity of board members to the appropriate limitations of their role. Subsequently, however, the board participated in the guidelines' development. Nevertheless, in its report to the administrator on the perchloroethylene addendum review, the board seemed to carefully limit the scope of its advice. The board's choosing to frame the animal evidence for perchloroethylene in such a way that a rote application of the guidelines would lead to a C classification rather than to argue that the agency should invoke the flexibility of the guidelines to arrive at the same conclusion may have reflected an assumption by the board that its influence on the agency's application of the guidelines was or should be limited. More recent communication between the agency and the SAB regarding perchloroethylene suggests a broader role for the board. Prompted by the administrator, the board was moved to comment not only on the agency's application of the guidelines but on the way in which the category assignment is used in the regulatory setting.

The board's movement into a more overtly science-policy arena such as that of risk characterization conceivably carries with it the danger of undermining the board's appearance of independence. But some opinion exists that providing advice to the agency on the policy implications of scientific judgments and the language used to communicate these judgments is not only an acceptable but a desirable SAB function. In Doull's opinion, "[The board] has realized that it isn't enough to assure that the quality of the science is good. Scientists can't just get by with saying something about the data, we also are obligated to say something about what the data means."[87] Doull does not believe that contributing an opinion about the latter is likely to threaten the board's independence. As he stated, "In the final analysis, [the board] just gives advice. We're trying to improve the *quality* of our advice."[88]

NOTES

The views presented in this chapter are those of the author and do not necessarily reflect those of the U.S. Environmental Protection Agency.

1. *Federal Register*, vol. 50, December 26, 1985, p. 52880.

2. *Chemical Regulation Reporter*, vol. 10, August 16, 1985, p. 512.

3. U.S. Environmental Protection Agency, *Health Assessment Document for Tetrachloroethylene*, Washington, D.C., July 1985.

4. Interview with Mark Greenburg, July 5, 1988, EPA.

5. *Federal Register*, vol. 51, September 24, 1986, pp. 33992–34003.

6. Interview with David Patrick, July 18, 1988, formerly of the EPA.

7. *Federal Register*, vol. 42, May 25, 1976, pp. 21402–5.

8. Ibid., p. 21404.

9. *Chemical Regulation Reporter*, vol. 4, September 12, 1980, p. 775.

10. U.S. Department of Health, Education, and Welfare, Public Health Service, National Institutes of Health, *Bioassay of Tetrachloroethylene for Possible Carcinogenicity*, DHEW Pub. no. (NIH)77–813, Bethesda, Maryland, 1985.

11. U.S. EPA Science Advisory Board, Subcommittee on Airborne Carcinogens, *Transcript*, September 4, 1980, p. 131 (hereinafter *Transcript*, 1980).

12. Ibid.

13. Ibid., p. 130.

14. Ibid., p. 137.

15. *Chemical Regulation Reporter*, vol. 4, September 12, 1980, p. 775. See also *Transcript*, 1980, p. 226.

16. *Chemical Regulation Reporter*, vol. 4, September 12, 1980, p. 775.

17. Ibid., p. 775.

18. *Transcript*, 1980, pp. 139–43.

19. Ibid., pp. 149–50.

20. Ibid., pp. 132–51.

21. Interview with Charles Hiremath, February 26, 1988.

22. *Transcript*, 1980, pp. 186–212.

23. *Chemical Regulation Reporter*, vol. 4, September 12, 1980, p. 775. See also *Transcript*, 1980, pp. 127–30.

24. *Chemical Regulation Reporter*, vol. 4, September 12, 1980, p. 775.

25. Ibid.

26. Interview with Mark Greenburg.

27. J. D. Graham, "The Failure of Agency-Forcing: The Regulation of Airborne Carcinogens under Section 112 of the Clean Air Act," *Duke Law Journal*, 1985, p. 113.

28. *IARC Monographs on the Evaluation of the Carcinogenic Risk of Chemicals to Humans*, Supplement 4, International Agency for Research on Cancer, Lyon, France, 1982, pp. 7–14.

29. U.S. EPA, *Health Assessment Document for Tetrachloroethylene (Perchloroethylene), Draft*, EPA–600/8–82–005, Washington, D.C., January 1982.

30. U.S. EPA Science Advisory Board, Environmental Health Committee, *Transcript*, September 28, 1982, pp. 187–94 (hereinafter *Transcript*, 1982).

31. Ibid., pp. 204–5.

32. Ibid., p. 204.

33. Ibid., p. 205.

34. Ibid., pp. 206–7.

35. *Chemical Regulation Reporter*, vol. 6, October 10, 1982, p. 780.

36. *Transcript*, 1982, p. 220.

37. U.S. EPA, *Health Assessment Document for Tetrachloroethylene (Perchloroethylene), Review Draft*, EPA–600/8–82–005 B. Washington, D.C., December 1983.

38. Ibid., p. 9–45.

39. *Federal Register*, vol. 49, March 21, 1984, pp. 10575–76.

40. Ibid., p. 10576.

41. *Transcript*, U.S. EPA Science Advisory Board, Environmental Health Committee, May 19, 1984, pp. 79, 84.

42. Ibid., pp. 36–37.

43. Ibid., p. 84. On June 4, 1984, a proposed recommended maximum contaminant level (RMCL) for perchloroethylene of zero was published in the *Federal Register*.

44. Ibid., p. 99.

45. Ibid., p. 21.

46. Letter from Herschel Griffin, Chairman, Environmental Health Committee, and Norton Nelson, Chairman, SAB Executive Committee, to William Ruckelshaus, Administrator, U.S. EPA, January 4, 1985.

47. *Federal Register*, vol. 49, November 23, 1984, pp. 46294–46300.

48. *Federal Register*, vol. 57, September 24, 1986, pp. 33992–34003.

49. Ibid., p. 33993.

50. Ibid., p. 33999.

51. U.S. Department of Health and Human Services, National Institutes of Health, National Toxicology Program, *NTP Technical Report on the Toxicology and Carcinogenesis Studies of Tetrachloroethylene in F344/N Rats and B6C3F1 Mice*, NTP TR 311, Research Triangle Park, North Carolina, August 1986.

52. Ibid., p. 2.

53. Ibid., pp. 14–15.

54. Memo from Elizabeth Anderson, Director, Office of Health and Environmental Assessment, to Joseph Cotruvo, Director, Criteria and Standards Division, John O'Con-

nor, Director, Strategies and Air Division, and Don Clay, Director, Office of Toxic Substances, U.S. EPA, September 29, 1985.

55. Ibid.

56. *Federal Register*, vol. 50, p. 47025. On June 4, 1984, a proposed recommended maximum contaminant level (RMCL) for perchloroethylene of zero was published in the *Federal Register*.

57. Ibid., pp. 52880–84.

58. Ibid., pp. 52881–82.

59. U.S. EPA, *Addendum to the Health Assessment Document for Tetrachloroethylene (Perchloroethylene), Review Draft*, Washington, D.C., April 1986.

60. Submittal from the Halogenated Solvents Industry Alliance to the U.S. EPA, April 18, 1986, Docket no. A–85–03, p. 2.

61. Ibid., pp. 2–3.

62. Ibid., p. 3.

63. Interview with Jerry Blancato, February 26, 1988.

64. Interview with Ernest McConnell, May 6, 1988.

65. Interview with Curtis Klaassen, May 31, 1988.

66. Letter from Rudolf Jaeger to John Doull, Chairman, SAB Halogenated Organics Subcommittee, May 19, 1986; emphasis in original.

67. Letter from Richard Griesemer, Chair, SAB Environmental Health Committee, and Norton Nelson, Chair, SAB Executive Committee, to Lee Thomas, Administrator, U.S. EPA, January 27, 1987.

68. Interview with John Doull, July 1, 1988.

69. Interview with Daniel Byrd, February 26, 1988.

70. U.S. EPA, *EPA Staff Comments on Issues Regarding the Carcinogenicity of Perchloroethylene (Perc) Raised by the SAB*, July 30, 1987.

71. *Overall Evaluations of Carcinogenicity: An Updating of IARC Monographs Volumes 1 to 42*, Supplement 7, International Agency for Research on Cancer, Lyon, France, 1987, pp. 29–32.

72. Interview with Penny Fenner-Crisp, April 1988.

73. Interview with Jeanette Wiltse, April 1988.

74. Letter from the Halogenated Solvents Industry Alliance to Lee Thomas, Administrator, U.S. EPA, June 1, 1987.

75. Letter from Lee Thomas, Administrator, USEPA, to Norton Nelson, Chair, SAB Executive Committee, August 3, 1987.

76. Letter from John Doull, Chair, Halogenated Organics Subcommittee, Richard Griesemer, Chair, Environmental Health Committee, and Norton Nelson, Chair, SAB Executive Committee, to Lee Thomas, Environmental Health Administrator, U.S. EPA, March 9, 1988.

77. Ibid., p. 2.

78. Ibid., p. 3.

79. Interview with John Doull.

80. Letter from John Doull, Richard Griesemer, and Norton Nelson to Lee Thomas.

81. Ibid., p. 4.

82. Interview with Penny Fenner-Crisp, emphasis added.

83. Interview with Jeanette Wiltse.

84. Interview with anonymous U.S. EPA official, Criteria and Standards Division, Office of Drinking Water, July 11, 1988.

85. Ibid. See also Natural Resources Defense Council, Inc. v. E.P.A., 824 F.2d 1211 (D.C. Circ. 1987).

86. See Terry F. Yosie, "Science and Sociology: The Transition to a Post-conservative Risk Assessment Era," Plenary Address before the Annual Meeting of the Society for Risk Analysis, Houston, Texas, November 2, 1987.

87. Interview with John Doull.

88. Ibid., emphasis added.

CHAPTER 7

FORMALDEHYDE

Susan W. Putnam

Formaldehyde has had a long and rather rocky history in the U. S. federal regulatory system. For almost ten years, since the Chemical Industry Institute of Toxicology (CIIT) released its interim bioassay results in 1979 indicating that formaldehyde inhalation leads to nasal cancer in rats,[1] government agencies have struggled with the issue of whether to compel reductions in human exposures to the chemical.

The controversy over formaldehyde stems primarily from its possible carcinogenicity to humans. Although there are positive animal data, the question of whether the chemical is a human carcinogen—especially at low doses—is still vigorously debated. The results of various quantitative risk assessments vary tremendously depending on what measures of exposure are used, which types of tumors are counted, and what modeling assumptions are employed. Scientists have not reached a consensus as to how the available data should be interpreted.

More recently, CIIT scientists have sought to clarify the controversy by undertaking an ambitious inquiry into the mechanisms of formaldehyde carcinogenesis. The most striking finding from this ongoing inquiry concerns the relationship between the amount of formaldehyde inhaled ("administered dose") and the amount of formaldehyde that reaches DNA in target cells in the nasal cavity ("delivered dose"). Although risk assessors typically assume a proportional relationship between administered dose and delivered dose, the new data from CIIT suggest that a nonlinear relationship may be operating at lower doses. The implication for human risk assessment is that inhalation of small doses of formaldehyde may not be as harmful as agency risk assessors have predicted.

Yet the new CIIT data still have not produced a consensus. Agency risk assessors have been reluctant to use the new findings in their risk assessments. They fear that these findings may not be reliable for use in human risk assessment, and they know that the data will produce smaller risk estimates (i.e., be less "conservative") in the low-dose region where regulatory policy is made. CIIT scientists have been frustrated by this reluctance and have continued to push forward in an effort to make their case ever more convincing.

The scientific controversies about formaldehyde highlight several important issues central to the use of science in the regulatory arena. The first of these issues concerns the influence of "settled" versus "frontier" science in regulatory decision making. Although the bioassay results were quite readily accepted by both scientists and regulators as indicating formaldehyde's carcinogenicity in rats, the later mechanistic or pharmacokinetic studies have provoked heated debates. The field of pharmacokinetics is considered by some to be on the scientific "frontier" and not yet appropriate for incorporation into regulatory decision making. Should these "frontier" data be incorporated into the risk-assessment and regulatory processes? Who decides when they have become "settled" and thus acceptable?

Related to this issue lies the search for scientific "consensus." Should regulatory decisions be based on data for which there is still scientific dispute? Although there have been several formal attempts to reach scientific consensus regarding the formaldehyde data, many uncertainties and controversies remain. This problem is especially apparent with the more recent CIIT data, particularly with the delivered-dose data.

A third area of importance concerns the proper role of CIIT in the regulatory process. With its release of the bioassay results in 1979, CIIT was immediately thrust into the midst of the formaldehyde story. An industry-sponsored research institution (whose data had been developed largely for nonregulatory purposes) became a central player in the formaldehyde controversy. Hence a central issue arose: should CIIT simply let its information speak for itself in the scientific literature, or should the organization use its scientific credibility more explicitly in the political process? How should CIIT balance the regulatory interests of its sponsors with its need to remain relatively insulated from the political aspects of the regulatory process?

In conjunction with all of these issues concerning formaldehyde lies the role of scientific peer review and advisory boards in the regulatory process. When there is uncertainty about how data should be used or dispute over their acceptability, what is the proper role of scientific reviewers in the decision-making process? Should it be the anonymous reviewers in the scientific community who decide implicitly what are appropriate data for risk assessments? Alternatively, should scientific advisory bodies be the arbiters in the risk-assessment process? What is the appropriate function of a scientific advisory board in such a case as the formaldehyde one, where the data are controversial and investigators cannot reach consensus as to what the findings show?

Although the scientific debate over formaldehyde surfaced in several federal agencies,[2] this chapter concentrates on the more recent events occurring at the Environmental Protection Agency (EPA). In particular, it examines the role of

CIIT and the EPA's Scientific Advisory Board (SAB) in the regulatory process for formaldehyde.

WHY FORMALDEHYDE IS IMPORTANT

Formaldehyde is a simple compound comprised of carbon, hydrogen, and oxygen. It is widely used in many industries, including the manufacture of plywood and particleboard, insulating materials, permanent-press fabrics, cosmetics, drugs, and protective coatings. The direct manufacture of formaldehyde has been estimated to constitute a $400-million business, while products containing the chemical represent approximately 8 percent of the U.S. gross national product.[3]

Because of these multiple and varied uses, many people are exposed to formaldehyde. The pathways of exposure are numerous, including food and tobacco smoke as well as occupational and environmental contacts. The most ubiquitous exposure concerns that of ambient air. In outside environments the ambient air may contain several parts per billion (ppb) of formaldehyde, mostly from automobiles and other sources of combustion, while indoor air may contain even higher levels due to such building materials as plywood, particleboard, or urea-formaldehyde foam insulation (UFFI). The highest levels of exposure to formaldehyde, however, occur in workplaces where the chemical is either used or produced, or in pathology laboratories where formaldehyde is used in preserving body tissues.

Formaldehyde has long been known as an irritant, capable of producing strong allergic sensitivities in humans, such as watery eyes, skin rashes, throat problems, and general discomfort. Because of these irritant effects, the chemical has been subject to federal regulation for many years, both as a manufacturing chemical in the workplace under the Occupational Safety and Health Administration (OSHA) and as a component of UFFI products in the home under the Consumer Products Safety Commission (CPSC). But formaldehyde was not given much consideration as a potential carcinogen before the CIIT results became available.

CIIT AND FORMALDEHYDE

CIIT's initial involvement with formaldehyde was relatively routine. Formaldehyde was selected for study because of its potential toxicity for humans, its wide usage, and its large human exposure. Major toxicity studies on this substance had not previously been done.

Because CIIT is an industry-sponsored organization, the issue of the institute's independence and objectivity is often raised by critics. But, as stated by Robert Neal, former president of CIIT, although industry sponsors make suggestions each year as to what substances CIIT should investigate, the ultimate research decisions are left up to the scientists and the management of the institute.[4] Also, as others would argue, studies by "disinterested" organizations are not necessarily "holier" than others. Scientists and regulators should not ignore studies by "interested" organizations, such as CIIT. The "color" of the research dollar should not matter, particularly in an era of decreasing funds available for scientific research.[5] In this case there was no industry pressure to specifically look at formaldehyde.

When the preliminary bioassay results came out in 1979, the chemical industry was as surprised by the results as everyone else. There had been no prior indication of what the results would be; most industry officials learned the news through a general press release. Although CIIT receives its funding in part from the chemical industry, it holds a disclosure agreement whereby the chemical companies do not have prior access to knowledge of or results from the experiments, nor are they allowed to monitor ongoing studies.

The formaldehyde results hit the chemical industry hard. This was the "first example of one of its [industry's] big oxes being gored by a creature of its own creation"[6] and represented the "coming of age" of CIIT.[7] Many of the chemical companies were very upset by the results, and one company even hired a "scientist" to investigate the data, who was very critical of both CIIT and the study.[8] Yet CIIT's leadership and board of directors stood firmly behind the scientific findings, thereby enhancing the organization's reputation for scientific integrity. The uproar within certain segments of the industry died down after several months, and the sponsors made no effort to muzzle CIIT or influence any further work on formaldehyde.

The institute continued its work with formaldehyde during the mid–1980s, again coming to the front with the 1984 Casanova-Schmitz study on delivered dose that became the subject of the special review panel recommended by the EPA's Science Advisory Board.[9] Although CIIT's interest in the chemical has tapered off in recent years (whereas CIIT initially allocated about 10 percent of its budget to formaldehyde research, this percentage has substantially declined over the years), it still maintains a research commitment to the chemical and is currently involved in a major primate study of formaldehyde.

REACTIONS TO THE CIIT BIOASSAY

The CIIT bioassay results stirred up regulatory interest and controversy at several federal agencies, including OSHA and CPSC, as well as EPA. The possible human carcinogenicity of formaldehyde triggered attention under several of the agency mandates and had strong implications for regulatory attention. Also, current with the release of the CIIT results was the development of "generic" cancer policy guidelines for identifying suspected carcinogens. Under these new guidelines regulators were instructed to consider positive results from animal experiments as establishing a cancer risk to humans.[10] As soon as these guidelines were drawn up, here was a chemical with which to test them.

Although this chapter focuses on the EPA and its consideration of formaldehyde, it is useful to provide a brief synopsis of what happened at CPSC and

OSHA to help illustrate the uncertainties and controversies involved with this chemical. Each of these regulatory agencies reacted to the CIIT results in a different manner, and each followed its own course of action (or nonaction) over the early 1980s.

CPSC took the most aggressive path of the three agencies. It was the first federal agency to take formal action against formaldehyde; after several initial proceedings it issued a total ban on residential uses of urea-formaldehyde foam insulation in 1982. It based this action firmly on the CIIT study results and also on the results of the 1980 report of the Federal Panel on Formaldehyde, which concluded that formaldehyde should be presumed to pose a risk of cancer to humans.[11] Although this regulatory action of the CPSC was later vacated by the Fifth Circuit Court of Appeals, it did represent a strong belief in the carcinogenic risks of formaldehyde and a regulatory commitment to prevent these risks.

In contrast to the CPSC, OSHA took a much less aggressive course regarding formaldehyde. Leadership at OSHA in the early 1980s was heavily influenced by newly elected President Reagan's policy of "regulatory relief," and the agency was hesitant to initiate any rulemaking proceedings on formaldehyde. The agency did initiate a new rulemaking on the chemical in 1985, but the justification for it was based entirely on formaldehyde's irritant effects rather than on its cancer risk. OSHA was not convinced that the CIIT data suggested a serious carcinogenic risk to humans. It was not until the fall of 1987 that the agency incorporated cancer risk as a rationale in a rulemaking on the chemical, lowering the workplace standard from a time-weighted average exposure of 3 parts per million (ppm) to one of 1 ppm.[12]

At EPA, the reaction toward the CIIT data took on the appearance of a seesaw. The agency's handling of the chemical was clouded in controversy and confusion, mainly as a result of the opposing policies of two of the agency's administrators, Anne Gorsuch and William Ruckelshaus. Each of these administrators, like the leadership at OSHA, adhered to the general approach of "regulatory relief" of the early 1980s, but each interpreted the scientific data on formaldehyde in a different light, and the two took nearly opposite actions regarding the issue.

EPA's regulatory authority over formaldehyde stems from the Toxic Substances Control Act of 1976 (TSCA). In particular, the controversy over action regarding formaldehyde concerns Section 4(f) of TSCA, which states that when a chemical presents a "significant risk of serious or widespread harm to human beings from cancer, gene mutations, or birth defects," the administrator of EPA shall initiate appropriate action to prevent or reduce to a sufficient extent such risk, or shall publish in the *Federal Register* a finding that such risk is not unreasonable.[13] EPA has proposed that either large individual risks ("serious" risks) or large numbers of exposed persons ("widespread" risks) will justify priority designation under the section.[14]

Under Anne Gorsuch the EPA decided not to list formaldehyde as a priority chemical under Section 4(f) of TSCA. This decision was supported by a sixteen-page memorandum from John Todhunter (the acting assistant administrator for the Office of Pesticides and Toxic Substances) stating that although formaldehyde might pose some risk of concern to humans, the risk was not sufficiently compelling to merit a 4(f) designation.[15] This memorandum was widely criticized on many fronts; cited reasons included improper industry influence, defective legal interpretations, and poor scientific analysis.[16] The EPA was also criticized for not seeking peer review regarding Todhunter's analysis.

In the spring of 1983 Anne Gorsuch resigned her post as administrator of the EPA, and William Ruckelshaus took over the agency. He was immediately faced with a legal suit from the Natural Resources Defense Council (NRDC) and the American Public Health Association (APHA) concerning EPA's decision not to consider formaldehyde under Section 4(f).[17] After spending several months trying to negotiate a settlement of the suit, Ruckelshaus announced that EPA would reconsider the 4(f) decision on formaldehyde.[18]

EPA decided to apply 4(f) consideration to two formaldehyde exposure groups: garment workers involved in the manufacture of apparel from fabrics treated with formaldehyde resins, and residents of conventional and manufactured (mobile) homes that contained construction materials in which certain formaldehyde resins are used (such as plywood or particleboard). The agency justified its decision for this policy reversal on the basis of quantitative risk estimates incorporating the CIIT data.

By 1985 EPA had developed its risk estimates and drawn up a formal quantitative risk assessment. It is here that the Scientific Advisory Board (SAB) enters the story. Before examining the role of the SAB, however, it is important to take a brief look at the formaldehyde data and some of the uncertainties lying therein. It is these uncertainties that underlie much of the importance that the Office of Pesticides and Toxic Substances placed on SAB review.

KEY SCIENTIFIC ISSUES

Much of the controversy over formaldehyde focused on the issue of whether the chemical should definitively be considered a human carcinogen. The CIIT study results indicated that at 14.3 ppm (the highest level of exposure tested in the study), formaldehyde was clearly carcinogenic to the Fischer 344 rat. Fifty percent of the rats developed squamous cell nasal tumors at this exposure level. But at the next lower level tested—5.6 ppm—only 1 percent of the rats developed tumors, and these results were not statistically significant. At the lowest level tested (2 ppm), there were no tumors observed in the rats. Also, for B6C3F1 mice tested at the same exposure levels, there were no statistically significant tumor results at any of the exposures (see Table 7.1).

These results left many questions unanswered. One important issue concerned the nonlinearity of the dose-response curve for formaldehyde in rats. The exposure appeared to have a proportionately much greater effect at the highest dose; that is, while the dose at 14.3 ppm was approximately two and a half times that at the 5.6 ppm level, the number of cancers differed by a factor of fifty.

Table 7.1
Adjusted Incidence of Squamous Cell Carcinoma of the Nasal Cavity:
Formaldehyde Inhalation for 24 Months

Formaldehyde Concentration (ppm)[a]	Number of Tumors/Animals at Risk[b]	
	Rats	Mice
0	0/208 (0%)	0/72 (0%)
2	0/210 (0%)	0/64 (0%)
6	2/210 (1%)	0/73 (0%)
15	103/206 (50%)	2/60 (3.3%)

Source: James E. Gibson, ed., *Formaldehyde Toxicity*, Hemisphere Publishing Corp., Washington, D.C.,1983, p. 297.

Note: Inhalation was for 6 hours/day, 5 days/week. The study was initiated with 960 Fischer 344 rats and 960 B6C3F1 mice, evenly divided by sex into treatment groups.

[a]Target concentrations. Actual average measured concentrations were 0, 2.0, 5.6, and 14.3 ppm.

[b]Actual number of animals exposed to formaldehyde up to, and including, the interval when the first squamous cell carcinomas were found (11–12 months for rats; 23–24 months for mice).

This difference raised the question of whether the dose-response relationship should be interpreted as linear at low doses (as risk assessors often do). If the relationship were considered to be truly nonlinear, then there might be a threshold for the chemical below which there would not be any adverse health effects. But if the relationship were to be considered linear, then there might be some effects even at very low levels of exposure.

This issue of linearity raised other scientific questions as well. Given this apparent curve for rats, what was the appropriate curve for humans? Do humans react in a similar manner to rats, or are there species-specific mechanisms at play? Particularly since there were very few tumors in the mice that were given the same formaldehyde exposures (and also few tumors in hamsters exposed in similar experiments), the extrapolation across species was left undefined. Also, given that the tumors developed at high levels of exposure—much higher levels than humans are currently exposed to—how should these results be extrapolated to lower-level exposures occurring in the workplace or environment? There were also uncertainties concerning mechanistic issues, such as whether the benign tumors (i.e., the papillomas and adenomas resulting from formaldehyde exposure) progressed into malignant tumors.

Many of these issues arise when one is deciding how to best fit the observed animal carcinogenicity results to a mathematical model predicting cancer risk for humans. There were several available models that fit the CIIT data fairly well but were based on different assumptions and shapes of the dose-response relationship. Different modeling processes produced risk estimates that differed from each other by several orders of magnitude. Also, the maximum likelihood estimates (MLEs) of risk and the upper confidence limits (UCLs) on risk were strikingly different. In short, there were many uncertainties associated with how to use the animal data to quantify the risk to humans at low exposure levels.

In conjunction with the animal bioassay data, there were also epidemiologic studies available that had examined the possible carcinogenic effects of formaldehyde. These studies, however, suffered from serious drawbacks that led to their results being far from conclusive. Most of these studies had been conducted in workplace environments, where workers were concurrently exposed to a myriad of substances. With these other exposures working as possible confounders, it was difficult to separate the effects of the formaldehyde exposure from those of the various other substances. In many cases, even smoking—a major confounder in that cigarette smoke both contains formaldehyde itself and also independently causes various problems for the respiratory system—was not controlled for confounding effects. Also, the exposures to formaldehyde that workers received were much lower than what the rats had received in the animal experiments. These different exposure levels made it difficult to extrapolate from the rats' situation to what the humans were experiencing. Finally, because the sample populations for the studies were relatively small and the subjects were not followed all the way to their deaths, the number of cancers that were expected to show up was very small. To be able to definitively observe excess cancers in these groups required sample sizes in the hundreds of thousands, as opposed to the several thousands of subjects that these studies generally involved. In short, the epidemiologic studies did not provide conclusive results on the question of human cancers from formaldehyde exposure; instead, they magnified the uncertainties involved and added more room for controversy.

ELUSIVE SCIENTIFIC CONSENSUS

While regulators at various agencies were trying to discern what the CIIT results really meant, there were several federal scientific panels analyzing the data as well. These panels were an effort to synthesize the information on formaldehyde, which could then be used to help solve the regulatory dilemmas facing the agencies.

The first such panel was the Federal Panel on Formaldehyde, formed under the auspices of the National Toxicology Program (at the request of CPSC), which was a group of scientists drawn from several of the major federal research institutions. In considering the various literature and data available on formaldehyde, the panel concluded that formaldehyde should be presumed to pose a carcinogenic risk to humans.[19] However, the panel did not carry out a quantitative risk assessment on the chemical or provide a numerical estimate of excess cases of cancer that could be expected from human exposure to formaldehyde. The resultant report, released in 1980, was criticized both for its lack of quantification and its apparent lack of substantive reasoning to support its conclusions, and it became less and less relevant to the debate about formaldehyde as agencies moved closer to the point of decision making.[20]

The second major group effort for a federal scientific review of the formaldehyde data occurred in the fall of 1983 with the Consensus Workshop on Formaldehyde. The workshop was organized by the EPA and the National Center for Toxicological Research and included about sixty experts on formaldehyde from various institutional settings. The participants were asked to comment on issues ranging from the implications of the existing data to where the needs for further studies lay. The goal was to present a less adversarial forum where scientists could reach consensus on the important issues concerning the chemical. However, from the standpoint of regulatory decision making, the workshop did not provide the hoped-for answers, as it failed to resolve many of the key technical disagreements at the heart of the formaldehyde controversy.[21]

Both the Federal Panel on Formaldehyde and the Consensus Workshop on Formaldehyde had been created in the hope of settling the scientific controversies for formaldehyde. The groups did work with agencies to develop risk assessments and policies for the chemical, but many of the scientific uncertainties remained.

At the EPA, after the agency had decided to reconsider formaldehyde as a 4(f) chemical, the Office of Pesticides and Toxic Substances (OPTS) employed the analyses of both the Federal Panel on Formaldehyde and the Consensus Workshop on Formaldehyde in developing its cancer risk assessment for the chemical. However, there were many questions left unanswered concerning the science on formaldehyde. The agency decided to seek additional review of these scientific issues from its advisory board, the SAB.

THE 1985 EPA QUANTITATIVE RISK ASSESSMENT

By 1985 the Office of Pesticides and Toxic Substances at EPA had developed a quantitative risk assessment for the health effects of formaldehyde. It had also narrowed its exposures of concern to those for garment workers and mobile-home residents. (It later passed its interest in garment workers over to OSHA, as it was authorized to do under TSCA.)

The risk-assessment document, entitled *Preliminary Assessment of Health Risks to Garment Workers and Certain Home Residents from Exposure to Formaldehyde,* included analyses of the various scientific issues pertaining to formaldehyde. Much of the document concerned the carcinogenic effects of the chemical, incorporating data from animal bioassays, epidemiology, metabolism and pharmacokinetic studies, and structure-activity relationships. Other sections of the document covered noncarcinogenic effects, exposure assessment, estimates of cancer and noncancer risks, and the characterization of the risk. The data in the document included the results of several animal bioassays—studies done by Albert and colleagues in 1982 and Tobe and colleagues in 1985, as well as the final results of the CIIT study published by Kerns and colleagues in 1983—and numerous epidemiologic studies of occupational populations.[22]

The document included numerical risk estimates for cancer from exposure to formaldehyde. The agency used the mùltistage model—based on the concept that the carcinogenic process usually involves multiple stages—and calculated both maximum likelihood estimates of risk and 95 percent upper confidence limits (UCLs). Use of the UCL from the multistage model resulted in a rather conservative estimate, in the sense of tending to overestimate the risks from chemical exposure rather than to underestimate them. Given the uncertainty involved in these risk estimates, the agency chose to err on the side of conservatism. From the analysis of the available data and modeling processes, the authors of the document concluded that formaldehyde was a carcinogen for rats and that the epidemiologic data indicated that the chemical "may" be a human carcinogen. Due to confounding exposures in the epidemiologic studies, there was not sufficient evidence to positively conclude that exposure to formaldehyde posed a carcinogenic risk to humans, but these studies were interpreted as suggesting that the chemical might be a human carcinogen.

Based upon the weight of the evidence, the agency then classified formaldehyde as a "Group B1—Probable Human Carcinogen" under its proposed Cancer Risk Assessment Guidelines. The B1 classification meant that there was limited evidence of carcinogenicity in humans from epidemiologic studies and sufficient evidence of carcinogenicity from animal studies. (This classification system ranges from A to D, where A represents a human carcinogen, B represents a probable human carcinogen, C represents a possible human carcinogen, and D represents a substance that is not classified because of inadequate animal evidence of carcinogenicity. Within the B category, there are two levels of classification, depending primarily on the degree of epidemiologic evidence available.)[23] The risk-assessment document also reported on various noncarcinogenic health effects from exposure to formaldehyde, but it was the carcinogenic evaluation that promised to influence the agency's regulatory position on chronic low-level exposures in the environment.

SAB REVIEW

The Decision to Seek SAB Review

After developing its risk-assessment document for formaldehyde, the OPTS subjected it to numerous informal reviews, both from inside and outside the agency.[24] OPTS also decided to seek additional, more formal review from the scientists on the Science Advisory Board. The SAB had been established as a public advisory group providing extramural scientific information and advice to the administrator and other officials at the EPA. The board was structured to provide a "balanced expert assessment of scientific matters related to problems facing the agency"[25] and to "identify areas where the scientific basis for decision making can be improved."[26] Under TSCA, SAB review of chemical analyses is not required by legislative mandate. OPTS made a voluntary decision to send the formaldehyde risk assessment to the SAB for review. In fact, this was the first time that OPTS was subjecting one of its risk-assessment documents to SAB review.[27]

There were still many uncertainties about formaldehyde, and OPTS wanted the SAB's opinion on whether it had laid out the scientific issues in a balanced and appropriate manner. The agency was still suffering the repercussions of the Todhunter memorandum fiasco, and it did not want to receive criticism again for failing to get SAB review. Also, OPTS was concerned about whether it had included all the appropriate available data. As Jack Moore, assistant administrator of OPTS, stated at the SAB meeting: "I want to make sure that this document is an accurate reflection of what is known, so that if this is going to be the basis of the Agency's decision from a risk management standpoint of what needs to be done, that we're working from the appropriate deck of cards."[28] Since this document was potentially going to be used by other EPA offices outside of OPTS,[29] there were agencywide incentives for SAB review of the data. Finally, the agency was concurrently revising its guidelines for carcinogen risk assessment, and OPTS wanted assurance that it was properly following the new procedures in the case of formaldehyde.

Formaldehyde had already been in the regulatory arena for many years and had built up strong constituencies on several sides of the regulatory battlefield. No matter what the agency decided to do, it would inevitably receive strong opposition and much attention. Opposing groups, such as the proindustry Formaldehyde Institute or the proenvironment Natural Resources Defense Council, had many resources and emotional investments tied to formaldehyde and were sure to oppose an agency decision that they considered to be not in their favor. Given formaldehyde's colorful history and the high "flack" number associated with it—that is, how much attention and opposition the agency would receive for its analysis and resultant policy—SAB review seemed to be the most logical step for OPTS to take.[30]

OPTS staff also saw the SAB review as a source of legitimization for their analysis. A negative review would allow them to reevaluate the issues of concern; a positive review would, in essence, put a "seal of approval" on their work.[31] It was also perceived that SAB review would legitimize not only the agency's scientific analysis and risk assessment but any future regulatory decisions as well.[32]

The SAB Meeting

In June 1985 OPTS presented its preliminary draft of the formaldehyde risk assessment to the Environmental Health Committee (EHC) of the SAB. This committee, one of several standing committees of the SAB, was comprised of nine scientists from various disciplines of science, including toxicology, epidemiology, pathology, and biostatistics. The scientists had been sent a copy of the risk-assessment document ahead of time, and the EHC members came to the public meeting in late June armed with prepared comments and ready to evaluate OPTS's work. There were also several ad hoc special consultants from particular areas of scientific interest invited to the meeting, as well as representatives from OPTS, SAB staff members Dan Byrd and Terry Yosie, and other interested parties (see the appendix at the end of the chapter).

Along with presenting the risk-assessment document to the EHC, OPTS had given the committee members a list of seven specific questions to address in their review:

1. Does the risk assessment appropriately utilize rat nasal squamous cell carcinomas and polypoid adenomas for hazard identification? Have the uncertainties in the tumor responses been adequately characterized?

2. Has the risk assessment conveyed to the reader the inherent variability in the risk extrapolated from the rat carcinoma data?

3. Has OPTS adequately extracted from the existing epidemiology studies the useful qualitative and quantitative information relevant to the assessment of formaldehyde carcinogenicity?

4. Has OPTS handled the area of nasal "physiological" barriers in an evenhanded manner? Can the data be better utilized?

5. How can one realistically begin to incorporate relevant kinetic information into quantitative cancer risk assessments?

6. Do ranges of "margin-of-safety" for noncarcinogenic endpoints adequately express the degree of concern (risk) for potential human exposures?

7. What guidance can be given concerning the evaluation of risks from carcinogenic and noncarcinogenic endpoints following expected human exposures to formaldehyde, given that it is a naturally occurring substance in the human body?[33]

These questions helped to guide the EHC review, and they focused the committee members' attention on particular issues. Also, in their written comments presented at the meeting, several of the committee members concentrated on those specific questions addressing the areas of science with which they were the most familiar.

The EHC meeting began with several presentations from industry scientists representing the Formaldehyde Institute. The Formaldehyde Institute took the position (which it still holds today) that there is no substantive evidence linking formaldehyde to cancer in humans,[34] and this position was reflected in its criticisms of the EPA risk-assessment document. Maureen O'Berg, an epidemiologist with Du Pont who addressed the epidemiology section of the document, expressed particular concern about the "selective citing" of positive findings, the inclusion of and reliance on findings that were neither published nor available in written form, the failure to fully address some important epidemiologic issues, and the document's general conclusions. Neil Krivanek, a toxicologist from Du Pont's Haskell Laboratory, emphasized the lack of evidence that formaldehyde produces tumors at distant sites and said that therefore the document should predict human effects only at the site of contact. He also stressed the nonlinearity of the dose-response curve in rats and the new data on DNA cross-links, suggesting that these data should be factored into the risk assessment. Another toxicologist, John Clary from Celanese Corporation, brought out the importance of the modeling process for risk assessment and argued that all the data, not just the rat data, should be incorporated into the risk calculations. Harold Imbus, an occupational physician formerly with Burlington Industries, was concerned about the importance of endogenous sources of formaldehyde and other exogenous sources and confounders. Finally, Kip Howlett from Georgia-Pacific discussed formaldehyde levels in wood products and the current decrease in formaldehyde emissions in manufactured and conventional homes. Overall, the presentations emphasized the lack of significant evidence indicting formaldehyde as a human carcinogen.[35]

After the individual presentations, the meeting was opened for general discussion by the EHC members to address issues in the document of special concern to them and then to try to answer the seven questions posed by OPTS. The committee's reaction to the document was largely one of general agreement and approval. Much of the day's discussion—as well as many of the committee members' prepared comments—focused on editorial details or specific technical issues. However, there were several significant areas in which the EHC found fault with the document and recommended that OPTS reconsider its analysis.

The first major issue concerned the document's review of the relevant epidemiology. The primary problem with this section was the way in which the epidemiologic studies were presented. There was some confusion over what the studies showed, and committee members wanted a clearer delineation—preferably in tabular form—of the various study characteristics (exposures, latency periods, and confounders) and results to enable a comparative analysis. In addition, it was recommended that OPTS weigh the studies according to how meaningful or powerful they were. The committee also suggested that the document present a more evenhanded discussion of the epidemiologic data, incorporating negative studies as well as positive ones and presenting "persuasive scientific arguments rather than those that seem to be based solely on the basis of an advocacy position."[36] There was also a recommendation that the document include more actual analysis of the studies as well as descriptions of them, particularly to illuminate the issues for readers who might not be experts in epidemiology. Many of these criticisms were fairly editorial and readily correctable, however, and the committee's general view was neatly summed up by EHC member Daniel Menzel:

It isn't such a bad document. I mean we always come off sounding like we hate documents, but it isn't such a bad document. It just needs to be cleaned up, and this particular section [epidemiology] is very important. I think that the language [should] be delineated so those of us who are not experts in this particular area can readily evaluate the document because if we can't evaluate it, it means that a broader audience cannot evaluate it.[37]

A more fundamental concern expressed by the committee involved the conclusions that should be drawn from the epidemiology. The interpretation of the epidemiologic studies was important to how formaldehyde was to be categorized—as a B1 or a B2 carcinogen—which could then have an impact on the ultimate policy decisions regarding the chemical. Also, as this was one of the first chemical analyses to fall under the new proposed cancer guidelines, there was added concern that the appropriate classification be used. Menzel stated:

[Concerning] the conclusion that there was adequate evidence to support the [classification] criteria suggested in the document . . . I think it's very important in this new set of guidelines that the Agency is trying to conform to, that we're very careful about expressing our opinion as to whether we agree that the classification for a particular chemical fits that particular category.[38]

Several committee members suggested that the document lacked explanatory reasoning regarding EPA's conclusion that there is limited (versus inadequate) human evidence of carcinogenicity. They felt that there was insufficient substantiation for the classification of formaldehyde as a probable—rather than a possible—human carcinogen. Many of the studies had small sample sizes and did not have statistically significant results. However, it was agreed that there had been some findings of human cancer that could justify the given conclusion of "limited" evidence. As stated by Leslie Stayner, an epidemiologist with National Institute of Occupational Safety and Health (NIOSH):

So overall, there have been some findings. I think I would agree with the language of limited. They are limited findings. . . . Certainly, the issue is not settled, but there are some suggested findings of an excess of nasal cancer. There are some suggestive findings of lymphatic brain cancers and a few other cancers as well. So I think at this point, limited is an appropriate term.[39]

The second major issue was the treatment of different kinds of tumors in the modeling process. Should papillary and polypoid adenoma (benign) tumors be handled in the same way as the squamous cell (malignant) tumors? Do benign tumors progress into some kind of carcinoma? Should there be some sort of conversion ratio to incorporate benign tumors into the risk-assessment process? Both Richard Griesemer (EHC) and James Swenberg (CIIT) felt strongly that benign and malignant tumors are two different endpoints and should be evaluated separately. Swenberg also emphasized the need for a conversion ratio because "the laws and everything [concerning risk modeling] deal with risk to cancer. And it might be useful to try to use a multistage model or something like that with the conversion ratio, or use whatever the best range you can in there to get a better handle on this.''[40] The committee felt that the risk-assessment document had not appropriately considered and evaluated these issues, and it recommended that OPTS go back and reexamine the data concerning the different tumor types.

Finally, a third area where the committee raised concern involved the use of pharmacokinetic data. The field of pharmacokinetics traces the uptake, metabolism, and distribution of a chemical in the body, including its interaction with DNA in target cells. By studying DNA adducts, it is hoped that more can be learned about the dose-response process, target-site mechanisms, and the overall process of carcinogenesis. Pharmacokinetics was still a developing field, and although there were several studies coming out in this area (including an important one by Mercedes Casanova-Schmitz at CIIT), OPTS chose not to include these data in the risk assessment. Some scientists at EPA did recognize that ''this is the direction that we have to go,''[41] but they were critical of the recent studies that had been done. As indicated by the fifth question put forth to SAB, they were unsure as to how to handle these data in the risk-assessment document.

Several members of the EHC, however, felt strongly that OPTS should have included the pharmacokinetic data in the analysis. Daniel Menzel, in particular, emphasized the importance of the pharmacokinetic data, believing that they pointed in the direction of basic pathways that would hopefully in the long run illustrate commonality between the carcinogens, and, therefore simplify our overall task of deciding what the implications of a particular study with a particular chemical are for human health.[42] He later added that ''there is already an adequate body of knowledge that one could incorporate kinetic information that is available on formaldehyde into it. So my comment [in response to question five from OPTS] is that they ought to do it [incorporate the relevant kinetic information] and not [just] begin to do it.''[43]

Committee member Robert Tardiff endorsed this point and suggested that even though not all was understood about the pharmacokinetic data, in presenting an evenhanded risk assessment, the OPTS should use the mechanistic information for a ''parallel analysis to determine the extent to which there may be risks in the human population, and the extent to which they may not be there.''[44]

Scientists from CIIT, who had been doing the pharmacokinetic research on formaldehyde, were frustrated with the EPA's position as well. As Swenberg stated:

While it [EPA] has paid lip service to mechanistic research, it has in actuality totally excluded it when it comes to the actual risk assessment process, and we could have done this very same risk assessment back in 1980 or '81 with the bioassay data. And I find this particularly disturbing. We had a great deal of emphasis in recent meetings and documents from the Agency and other scientists suggesting that we start doing risk assessment, incorporating the science, incorporating molecular dosimetry, if you will, and it was just ignored.[45]

Swenberg felt that the EPA ''had blinders on'' regarding the mechanistic data, and that there were not people at the agency to whom he could communicate the importance of the data.[46] Bill Farland, acting director of the EPA's Office of Health and Environmental Assessment, later agreed with this criticism and admitted that the agency should have been more receptive to CIIT's concerns at this point.[47]

Because of the controversy surrounding the pharmacokinetic data and their potentially important role in the risk-assessment process, EHC recommended that a special independent peer review panel be formed to evaluate the data on formaldehyde pharmacokinetics. More specifically, the committee suggested that this panel examine the Casanova-Schmitz study from CIIT and the ensuing analyses. Menzel stated:

If there is so much controversy surrounding it [the pharmacokinetic information] that people are uneasy, then maybe the likes of a peer review panel of those two papers [the original Casanova-Schmitz paper and the Buck and Starr paper analyzing it], or those two papers plus any of the others that the authors or the staff of OPTS would suggest, would be appropriate. Maybe that's the way to really focus the issue, and therefore, move us off dead center, and incorporate this kind of an idea better into the methodology. That would then be serving not only EPA's long term needs [exploring a new science], but also their short-term needs with regard to this particular [Casanova-Schmitz] paper.[48]

This suggestion was seconded by other members of the EHC and also was readily accepted by several EPA staff members present as a viable approach to important data.

Although the EHC members had already touched on many of the issues relevant to the OPTS's concerns in the course of the day's discussion, the last part of the meeting was devoted to directly addressing the seven questions posed by the OPTS. Specific questions were directed at specific committee members, according to their expertise, and many of the members had prepared written responses to these topics. Many of these comments were then incorporated directly into the SAB's report to the OPTS on these issues.

The SAB Report

After the public meeting in June, the EHC staff wrote its report, incorporating the committee members' comments and recommendations. The report was then sent to the administrator of the EPA the following October. Although the report was largely pulled together by the EHC staff with the help of Richard Griesemer, chairman of the EHC, and Norton Nelson, chairman of the Executive Committee of the SAB, it was meant to be a collaborative effort with all the EHC members participating.

At the EHC meeting in June, all of the committee members submitted written comments from their initial independent reviews of the risk-assessment document to the EHC chairman. Also, members could submit additional comments after the meeting if they so chose. Then, with help from the SAB staff, all of the various written comments and discussions from the meeting were incorporated into a draft report. This report was subsequently sent back to the committee members for review and revision in an effort to ensure that all of their thoughts had been properly captured. The report then went to the SAB Executive Committee for another review before it was sent to the administrator. Not only was the final report sent to the administrator, but all of the committee members' comments were sent as well. Whereas the report was generally issue-oriented, many of the comments tended to be more editorial or concerned with specific technical issues. Finally, although the final report was not sent until October, an immediate letter dated July 2, 1985, was sent to OPTS to provide an initial draft response to the seven questions that OPTS had requested the EHC to address. The committee realized that EPA needed its response as quickly as possible in order to evaluate the advice, to decide whether or not to adopt it, and then to proceed in the analysis.[49]

The final report—in the form of a letter—was sent to EPA Administrator Lee Thomas in October 1985. Again, the response was generally favorable, commending OPTS on many issues, but also drawing attention to several areas of concern.[50] These areas were similar to those raised at the EHC meeting in June. The main issues of epidemiology and pharmacokinetics were discussed, as well as some other issues concerning the structural layout of the document and the use of inaccessible and secondary information sources. However, the letter was quite short—approximately two pages—and has been criticized as presenting only a very limited view of what came out of the EHC meeting and the whole SAB review process:

No letter accurately reflects what goes on at the meeting. Even the transcript doesn't reflect accurately what goes on at the meeting. That's because people are thinking the whole time—so you have a limited view of what comes out of the process. That's the problem with these letter reports—they're such limited views.[51]

In addition to the letter, the report also included the EHC response to the seven questions posed by OPTS. In this response report the EHC went into more depth and gave relatively detailed advice addressing these specific issues. It is interesting to note, however, that this response report was almost identical to the one sent to OPTS immediately after the June meeting,[52] raising a possible question as to the amount of input that the individual committee members did have into this part of the report after the meeting. One committee member felt that individual members did not have enough input into the process, but that this was mainly due to a lack of SAB staff involvement and resources. He further commented, however, that over the last several years the SAB staff has taken a stronger leadership role in the process, and that this problem has largely been corrected.[53]

The Pharmacokinetics Review Panel

As previously noted, the EHC and OPTS disagreed over the incorporation of pharmacokinetic data into the quantitative risk-assessment process. Also, EHC did not feel qualified to thoroughly address this issue by itself. Thus it was suggested at the EHC meeting that a special review panel be formed to evaluate the appropriate pharmacokinetic studies. The response report from the EHC to the administrator more formally reiterated this suggestion, stating as its response to question five that the "consensus of the Committee is that the document will not be scientifically adequate without an analysis of the pharmacokinetic information and appropriate modification, based on this analysis.[54] To deal with this problem, the report went on to say that "the Committee suggests that one option to resolve this deficiency [insufficient pharmacokinetic analysis] in the draft document is for EPA to convene a workshop specifically on the pharmacokinetics of formaldehyde for the purpose of providing an evaluation for OPTS."[55] The other option suggested by EHC in the response report was to use the reports already available in the literature and then address the quantitative implications for the assessment by sensitivity analysis.[56]

OPTS chose the first option recommended by the EHC. It contracted with Life Systems, Inc., a scientific consulting firm, to bring together a panel of seven scientists with expertise in metabolism, DNA adducts, and statistical modeling to evaluate the pharmacokinetic studies. OPTS wanted a panel that would provide an "independent, objective, expert analysis of the issue so that it may reconsider the appropriateness of using the data in its risk assessment."[57] The scientists on the panel were selected on the advice of various scientists at EPA, in conjunction with Life Systems, Inc., on the basis of their specific areas of expert knowledge. Also, there was a concerted effort to cover the appropriate range of disciplines, while choosing scientists who did not have particular industry or agency affiliations.[58] As a result, the selected panel was mainly comprised of scientists from academic institutions. OPTS was pleased with the qualifications of the selected panel and generally considered it to have been well constructed for its task.[59]

The pharmacokinetics panel convened in early December 1985 to consider whether the available pharmacokinetic studies provided information that was suitable for use in assessing the health risks of formaldehyde. Prior to the meeting the panel members were asked to independently evaluate the documents concerning the Casanova-Schmitz study and to determine whether the pharmacokinetic data were appropriate for use in the formaldehyde cancer risk assessment. Like the EHC review board, these experts were asked to consider a series of specific questions about the CIIT study. The questions concerned both methodology and process issues associated with the study, as well as the appropriateness of incorporating the study into the OPTS risk-assessment document.[60]

At the panel meeting several representatives from CIIT participated in an informal discussion with the expert reviewers. This discussion stemmed from the experts' requests for information and clarification concerning several points in the Casanova-Schmitz study. It also included an expanded question-and-answer session during which further details were provided on CIIT's prior, ongoing, and planned studies of formaldehyde.[61] The panel wanted to be sure that it had all the appropriate data.

The final report of the pharmacokinetics review panel was issued on January 2, 1986. The report expressly addressed the eight questions that the panel had been asked to consider. Overall, the panel was very critical of CIIT's work, finding fault both with several of the experimental procedures that were used and with potential analyses that were omitted. The report recognized the Casanova-Schmitz study as an important step toward attempting to assess the intracellular dose delivery of externally applied formaldehyde and agreed that these efforts should be continued toward the ultimate goal of improving the assessment of risk. However, the panel concluded that at its present level of development and validation, the study did not represent an adequate basis for quantitative risk assessment. This conclusion was based on arguments that (1) experimental methodologies and assumptions had not been properly validated; (2) intracellular proteins rather than DNA should have been used as the target; and (3) acute exposures might not have been relevant to chronic exposures. In essence, the panel resolved that the study was an "important first step toward the introduction of intracellular dosimetry into the risk assessment process"[62] and that the continuation and extension of these investigations should be encouraged, but that it did not provide a basis for risk quantification.[63] The panel therefore supported the EPA's original position that the use of these data for risk-assessment purposes was "premature."[64]

Reactions to the Panel

The EPA was, in general, satisfied with the pharmacokinetics panel and its report. Frederick DiCarlo, the EPA scientist who was most involved with the pharmacokinetic data on formaldehyde, was "very pleased by the reports of the expert review panel on formaldehyde"[65] and felt that the panel had reinforced the EPA's decision not to use the pharmacokinetic data in the risk assessment. Even with the possibility that the special panel might have caused a delay in the agency's regulatory debate on formaldehyde, he believed that the expert review panel had been a worthwhile process, and that the panel's comments had been an important contribution to the revisions made to the 1985 risk-assessment document.[66] Other EPA staff concurred with DiCarlo's position and felt that the panel had conducted a very probing sort of analysis, had asked a lot of difficult questions, and, in general, had done a good job.[67]

Several SAB members who had originally recommended the formation of the panel were not in total agreement with the panel's report. They had hoped for a more positive response to the Casanova-Schmitz study. As fellow scientists, however, they respected the panel's work and attributed much of the panel's hesitancy to fully endorse the pharmacokinetic data to the inherent conservatism of scientists toward a subject that they may not totally understand.[68] As Menzel stated, although pharmacokinetics was an up-and-coming area of scientific interest (on the "frontier" of science), "it had not [yet] been anointed [and] was not ready for prime time."[69]

The most vehement opposition to the special panel's report, however, came from CIIT scientists. They considered many of the panel's concerns to be "without merit,"[70] and they "disagreed strongly"[71] with the panel's conclusions. They sent the EPA a highly critical statement of comments on the panel's final report. The document contained direct rebuttals to the panel's evaluations for each specific question and went into at least as much detail as the panel's original analysis, if not more. The CIIT scientists disagreed with all of the panel's final arguments and found its conclusions to be unsubstantiated.[72]

In particular, CIIT's position, as iterated by its then president, Robert Neal, was that the pharmacokinetics review panel was not qualified to make the judgments and criticisms that it had made. The panel did not contain scientists with the appropriate statistical backgrounds, nor enough people with the necessary knowledge of target-site dose or DNA binding mechanisms. It did not have the expertise to really understand the complicated methodologies that had been used in the study. Also, the Casanova-Schmitz study had already been peer reviewed in refereed scientific journals, and its scientific validity should not have been in question. The severe and deliberate criticism by the review panel was, according to Neal, neither substantiated nor scientifically justified.[73]

THE 1987 EPA RISK ASSESSMENT

After receiving the EHC report in the fall of 1985 and the pharmacokinetics panel report several months later, OPTS spent the next year reviewing the comments it had received on its preliminary risk-assessment document and reevaluating the areas of uncertainty. In April 1987 OPTS released a revised cancer risk-assessment document for formaldehyde.[74]

A comparison of the two versions of the risk assessment shows that one of the main differences was the inclusion of many new epidemiologic studies in the 1987 document, some of whose results had only become available since 1985. Several of these studies, such as the Blair study (NCI) and the Stayner study (NIOSH), were of particular interest because they involved relatively large study populations (26,581 and 11,030 workers respectively). These studies, along with the 1986 Vaughn study, were designed to detect moderate increases in formaldehyde-related risks, and they also addressed the possible effects of confounding factors on their results.

These studies, however, did not clear up the epidemiologic uncertainties. In fact, the NCI study became embroiled in such controversy that it may ultimately have served to confuse the issues further. The investigators of the study found no overall increase in cancer mortality or statistically significant increases in lung or upper respiratory tract cancers, and they subsequently concluded that there was "little evidence"[75] that formaldehyde caused cancer among exposed workers. However, the study's advisory panel strongly disagreed with this interpretation of the data and felt that the study did not resolve the issue of formaldehyde as a possible human carcinogen for the exposed workers.[76] This disagreement over the interpretation of the results subjected the study to much public scrutiny, and NCI may have lost some credibility over the fiasco. There were also allegations accusing NCI of allowing excessive industry influence on the conduct of the study and its ultimate release.[77]

The pluses or minuses of these new studies aside, they did not affect OPTS's overall classification of formaldehyde. The epidemiologic evidence was still seen as limited, and formaldehyde remained in group B1, probable human carcinogen.

Although the 1987 document did not encompass any major conclusions that differed from the 1985 version, it did make several structural changes that had been recommended by the EHC. The presentation of the epidemiology section was largely reworked so as to include a series of comprehensive tables incorporating all of the various studies. An effort was made to detail the individual study characteristics more clearly, and the studies were arranged so as to enable a comparison of various factors across the studies. Also, the risk-characterization section, considered by many to be the key section of the document, was reworked and moved from the end of the assessment to the beginning so as to attract more attention. Finally, even though the 1987 document did not change the agency's position on whether to combine benign and malignant tumors or whether to incorporate the Casanova-Schmitz pharmacokinetic results, it did include discussions of these issues and attempted to justify why the agency had not changed its position on them.

In short, although the risk-assessment document did not appear to change dramatically in response to SAB review, OPTS did at least pay attention to what the SAB committee had said. The epidemiology section was reworked, and the issues concerning pharmacokinetics and benign tumors were addressed. Even though OPTS had chosen not to include the pharmacokinetic data in the revised document, it had followed the EHC's recommendation to seek review of this issue from an independent panel. According to several OPTS staff members, although the EHC's comments had not been the "chief variable" in the 1987 revisions, they had been considered, along with the additional data that had become available, in the revising of the risk-assessment document.[78]

One could argue that the EPA took the easy way out in this situation and merely addressed certain issues superficially, never intending to change its assessment. But the EPA staff does not have to do what the SAB recommends. The board serves the agency in an advisory, not a compulsory, capacity. The EPA has the option of heeding, investigating, or ignoring SAB advice as it sees fit. If the agency staff ignores the SAB comments outright, it may then have to answer for this decision to regulatory opponents or federal judges. But the SAB advice is not the last word. If the agency staff investigates the recommendations and justifies why it chooses to oppose any of them, it has fulfilled its obligation. SAB is concerned about whether the agency evaluates the scientific data properly; it is not—or is not supposed to be—involved with the policy decisions that may arise from the science. As SAB member Herschel Griffin stated, "We're reviewing a document. We're not a jury here to decide whether formaldehyde is carcinogenic, but whether the data have been properly presented so that determinations can be made."[79]

Another criticism that could be leveled against the agency here concerns the special pharmacokinetics panel. Since the agency was essentially responsible for choosing the scientists to serve on the panel, it might be questioned how independent the panel really was. Especially in this case, where the panel reached the conclusion preferred by the agency, one's thoughts might drift toward the notion of collusion or influence. One must keep in mind, however, that the SAB situation is not unlike that of this panel. EPA, in essence, selects the scientists that will serve on the board, and it also allocates a budget for the SAB. Together with the SAB staff and committee members, it selects the substances that will receive SAB review. In this sense one may wonder how truly independent of EPA the SAB is.

CURRENT REGULATORY STATUS

After the 1987 risk assessment was released, OPTS activity on formaldehyde declined. The agency is still tracking the literature and monitoring ongoing studies concerning formaldehyde, but it is not devoting as many scientific resources to the chemical as it did in the mid–1980s. The focus of attention at EPA has shifted from risk assessment to risk management. The chemical has been turned over to the OPTS options committee to evaluate the risks and develop various policy options for formaldehyde.[80]

OPTS has decided to go with the 1987 risk assessment as the current working document. The agency is comfortable with the document as one that can be used in risk management. With the possible exception of CIIT's delivered-dose study of primates, there does not seem to be anything new coming "down the pipeline" that will have a major impact on the document, and the agency foresees that this risk assessment should stand up to scrutiny for at least a few years.[81]

Should the 1987 risk-assessment document go back to the SAB for review? EPA staff members believe that this is not necessary.[82] Since much of the document is the same as it was in 1985, they believe that another SAB review would be senseless. The parts of the document that did change, such as the epidemiology and risk-characterization sections, were developed in accordance with the EHC recommendations. The only reason they foresee that would justify sending the assessment back for rereview would be if some unexpected new data results were released or if the public were to make a big outcry in favor of another review.

Several other interest groups, however, do not share this view. The Formaldehyde Institute is pushing hard for the document to go back to the SAB, primarily because of the risk assessment's inclusion of the three new large epidemiologic studies (the Blair, Stayner, and Vaughn studies) that were not reviewed by the SAB in 1985.[83] In June 1987 a meeting was held between Chuck Elkins of OPTS and other EPA staff members, and Jack Murray, president of the Formaldehyde Institute, concerning the rereview of the risk assessment by the SAB. Whether anything will evolve from the meeting and the institute's push remains to be seen.

Proponents of pharmacokinetics also would seek a rereview on the grounds that the field of pharmacokinetics is continually evolving.[84] New data have come out in this area since 1987 (e.g., CIIT's delivered-dose information on primates), and it is important that the SAB ensure that the EPA stays abreast of new scientific developments. Also, those in favor of having repeated reviews to assure as much as possible that the risks are being assessed accurately would seek rereview. Environmentalists believe that the more times that the SAB can review a document, the better for those groups who might be regulated on the basis of risks.[85]

In light of the ever-increasing demands on the SAB's time and budget to review more and more substances, it is doubtful that the 1987 document will go back for another review. The SAB has given its seal—mostly of approval, in this case—and the EPA has moved formaldehyde into the next phase of attention.

THE ROLE OF SAB IN THE REGULATORY PROCESS

Upon a first glance it might appear that the SAB's role in the regulatory process at EPA for formaldehyde was small, perhaps even insignificant. In the two main areas of scientific concern over which EPA and SAB disagreed—pharmacokinetics and the relationship between benign and malignant tumors—EPA appeared to win out; that is, despite SAB's negative comments, EPA's analysis remained the same. The areas in which EPA did heed SAB's advice involved only structural or presentational issues, such as formulating epidemiologic data tables or rearranging certain sections of the document. In short, it might appear that SAB had some influence in the lower-stakes issues, but perhaps not in the larger scientific areas that had a lot riding on them.

The formaldehyde case suggests that SAB may not always be able to convert frontier science into settled science so that this information may be readily accepted by regulators and incorporated into risk assessments. Although several members of the SAB felt strongly about the importance and validity of CIIT's pharmacokinetic data—an area that was still considered to be on the frontier of science—EPA ultimately stayed with its original analysis omitting these data. Even with outspoken prodding from the SAB and CIIT, EPA chose the cautious route and shied away from the unsettled scientific data. SAB was unable in this case to convert pharmacokinetics into accepted regulatory science.

Yet the formaldehyde case does reveal the power of SAB as an institution. In addition to being another forum for scientific review for agency documents and issues, SAB review has taken on an identity all its own. Many regulators see SAB review as a must if a document is to be successfully accepted as appropriate science, particularly when there are many uncertainties involved or when the document will inevitably receive much public scrutiny. As one OPTS regulator stated:

A document that doesn't have [approval], that has not been reviewed by the SAB will be used in the regulatory process. [But] if that document has the "EPA-SAB seal of approval," if you will, then that document has much more . . . the agency is more willing to move forward. Decision-makers are much more willing to put themselves on the line, to take controversial stands, to impose regulations which cost money.[86]

The SAB has several important factors to its credit. The committees within the SAB are known for being comprised of prestigious scientists. The SAB meetings are held publicly and are open to anyone who wants to attend. Also, the review process, as a whole, is considered to be generally conducted in an aboveboard manner.

There have also been criticisms of the SAB process, however. One of the major criticisms concerns the issue of whether SAB review causes delays or generally slows down the regulatory process—particularly when the SAB is used as mediator between two EPA offices that cannot resolve their differences concerning a chemical. Other concerns question the quality of review that can be obtained from a very busy panel in a one-day review, and, given the prestige of the SAB, the possible harm that could come if the SAB's judgment were to be seriously in error.

In the formaldehyde case SAB's review was sought as a seal of approval. There were numerous uncertainties in the science, and, given the political and historical controversy surrounding formaldehyde, SAB review was seen as a way to validate the agency's risk assessment. Also, the issue of delay was not a major factor. The EHC responded to the OPTS questions right away, and even the special pharmacokinetics panel reported back to EPA within about six months of the SAB review meeting. The agency was also collecting additional epidemiologic data during this time, so the waiting time was not wasted.

The SAB review may not have been the chief variable in the revision process of the risk-assessment document, but evidence of the SAB recommendations is very apparent, particularly in the reworking of the epidemiology section. Also, the solicitation of the special panel on pharmacokinetics represented a fairly major undertaking in conjunction with the SAB's advice. Even if SAB may not have agreed with the conclusions of the pharmacokinetics panel, it could not fault the EPA for opting to side with the special panel after a careful review.

It has yet to be seen how the issue of SAB review as document or policy validation will play out in the formaldehyde case, as the EPA has not yet issued any regulatory policies on formaldehyde. One would surmise, however, that if the agency were to try to regulate the chemical as a probable human carcinogen, the SAB review of the risk assessment would become a central consideration in the political (and possibly legal) battles that would inevitably ensue.

CIIT AND THE GOVERNMENT

Given the critical role that CIIT's research could (and sometimes does) have on the regulatory process, one might ask whether government scientists should participate more directly in CIIT's activities. Should scientific experiments, such as those on formaldehyde, that might have a strong regulatory impact be reviewed in advance by the relevant agencies to allow suggestions for the most effective investigation for agency purposes?

CIIT's response is that regulatory agencies do not have the scientific expertise to review these types of experiments. Universities and research institutes (such as HEI and CIIT) are much better equipped to judge the validity of such experiments, and to rely on the regulatory agencies for guidance would be "counterproductive."[87] Also, in conducting scientific experiments, it is often better to ignore the regulatory issues and concentrate on searching for where the greatest human risks lie. In addition, CIIT has its own review board (as do many other research institutions), consisting of about twenty scientists from academic institutions, government laboratories, and private organizations, that reviews the scientific protocols, programs, and quality control at the institute every year. Moreover, CIIT believes that it should jealously guard its independence—both from its sponsors and from the government. The organization will not accept government grants, as it does not want to split its loyalties between industry and government. It also does not want the "strings" and loss of flexibility that often come with agency ties and funds.[88]

One may argue, nonetheless, that there should be more coordination between the agencies and CIIT to ensure the most efficient use of limited research funds and scientific talent. Particularly for studies commencing on already-hot regulatory topics, there should be more effort to marshal consensus on experimental techniques and to investigate major areas of uncertainty. EPA faces "alligators" no matter which way it goes; it is up to scientists to build a strong-enough case to support its entrance into new territories.[89] Also, if one believes that CIIT dove deeper into formaldehyde research in order to further explain its initial results, then perhaps there ought to have been some outside monitoring of the later experiments to ensure their scientific relevance and validity—at least in the eyes of government scientists. Perhaps agencies would be less inclined to resist inclusion of CIIT data if agency scientists had participated in protocol development. Also, although CIIT does not actually do risk assessments—they involve extrapolating judgments that are not specifically provable[90]—its data often become crucial components of agency risk assessments. Would risk assessors have more faith in these data, particularly those still on the scientific frontier, if they had more upfront input into the studies?

The arguments in this debate center around two main issues. The first focuses on the question of research for the sake of pure scientific investigation versus science as a contributor to the regulatory process. The second concerns the matter of how much influence in the regulatory process should be given to an independent research institution. The formaldehyde case suggests that there may still be some confusion about CIIT's role. Is high-quality science enough, or must the science also be perceived as relevant and compelling by the regulatory agencies? Should CIIT let its science speak for itself, or should it advocate use of its science in the regulatory process? Answers to these questions have important implications for the design of the organization.

OBSERVATIONS AND CONCLUSIONS

The actual impact of the SAB and its review in the formaldehyde case is somewhat clouded. The OPTS did heed several of the SAB's recommendations—particularly for small-stakes, presentational issues, such as the format of the epidemiology section—in the revisions of its risk-assessment analysis. The big-ticket issues that were on the frontier of science, however, did not change. OPTS followed the SAB's recommendation to have an independent panel review the pharmacokinetic data, but it did not heed the SAB's desire to have the data analyzed and incorporated within the document. Also, although OPTS staff stated that the SAB comments were taken into consideration in the revision process, they also implied that it was the new data that had become available in the interim between the risk assessments that had played the major role in the revisions.

SAB was not able to turn science on the ''frontier'' into ''settled'' science. Although several of the SAB scientists pushed hard for the pharmacokinetic data to be accepted as evidence, the data ultimately were not included in the risk assessment. The opinions and recommendations of the SAB members did not prove to be the bottom line for the acceptability of an area of frontier science for the OPTS.

Although in this case the EPA may appear to be more scientifically conservative than the SAB, being more hesitant to endorse issues on the scientific frontier, this may be more a result of its regulatory experience and omniscience of having to defend its analyses in court than an inherent unwillingness to incorporate something new. A regulatory position at the end of a limb is more easily severed than one more centrally positioned, and the EPA has been sued enough to quell its more tenuous claims about science.

The controversy over the formaldehyde data initiated by CIIT raises questions concerning the role of a private research institution in the regulatory process. With the release of the bioassay data, an industry-sponsored research organization emerged in the spotlight of the federal regulatory process. CIIT's leadership and scientists struggled with a difficult dilemma: how to remain independent of their sponsors and the government while pursuing work that was relevant to the risk-management decisions of industry and government. CIIT's handling of the formaldehyde issue has enhanced the institution's reputation for independence.

If the EPA had had a stake in the funding of the mechanistic data for formaldehyde, would that have made a difference in the regulatory decision-making process? Probably not. The agency had been hardened by the controversial history and lawsuits concerning formaldehyde, and it saw the conservative path as the only way it could go. In this case, as in others, "data alone are not enough to convince the agency to regulate"[91]—other political forces are involved, and it often requires a ricochet effect of issues bouncing between these forces for a decision to be made. Without a strong mover, as well as significant data, the regulatory process may become indefinitely mired in indecision or postponement.

The central role played by CIIT leads to the question of whether the federal agencies ought to be doing more mechanistic research themselves, so as not to have to rely on outside organizations for the relevant scientific data. Ideally, it would be more appropriate for the agencies to do their own research, but it is difficult for an organization to pursue both scientific and regulatory missions without confronting perceived conflicts of interest. Realistically, it seems doubtful that any of the agencies will be able to compete financially with private organizations to attract and hold the scientific talent that is needed to stay current. To be able to successfully build up a more substantial research program in the EPA or other agencies would require a major commitment from Congress to appropriate the necessary funds for investment in scientific expertise.

Finally, the EPA's lengthy and somewhat controversial investigation of formaldehyde raises the issue of whether the agency will ever make regulatory decisions about the chemical. Even after surpassing the hurdles of risk assessment and SAB review, the agency appears to be no closer to making a decision. The formaldehyde problem has provided a steady stream of funds for the OPTS over the 1980s. The political controversy and public awareness has guaranteed a certain level of funding for its investigators. Were the issue to be settled, the OPTS might lose an important appropriations trigger without having an equally prominent or "colorful" chemical to fill the financial void.

NOTES

1. Chemical Industry Institute of Toxicology, "Statement Concerning Research Findings," EPA Docket no. 11109, October 9, 1979.

2. For a detailed account of the regulatory history of formaldehyde, see John Graham, Laura Green, and Marc Roberts, *In Search of Safety: Chemicals and Cancer Risk*, Harvard University Press, Cambridge, Mass., 1988, Chapter 2.

3. Formaldehyde Institute, "Formaldehyde Backgrounder," Washington, D.C., undated.

4. Interview with Robert Neal, President of CIIT, Research Triangle Park, North Carolina, July 14, 1988.

5. Statement by Paul Portney, Director, Center for Risk Management, Resources for the Future, Washington, D.C., October 25–26, 1988 (Belmont Workshop).

6. Interview with Robert Neal.

7. Statement by Robert Neal, President of CIIT, Research Triangle Park, North Carolina, October 25–26, 1988 (Belmont Workshop).

8. Interview with Robert Neal; quotation marks emphasized by Dr. Neal.

9. M. Casanova-Schmitz, T.B. Starr, and H. Heck, "Differentiation Between Metabolic Incorporation and Covalent Binding on the Labeling of Macromolecules," *Toxicology and Applied Pharmacology*, 76, 1984, 26–44.

10. Graham, Green, and Roberts, *In Search of Safety*, p. 12.

11. "Report of the Federal Panel on Formaldehyde," *Environmental Health Perspectives*, vol. 43, 1982, p. 139.

12. *Federal Register*, vol. 52, no. 233, December 4, 1987, p. 46168.

13. U.S. Code, 2603, "Toxic Substances Control Act," Section 4(f), 1976.

14. Graham, Green, and Roberts, *In Search of Safety*, p. 29.

15. Ibid., p. 30.

16. Nicholas A. Ashford, C. William Ryan, and Charles C. Caldart, "A Hard Look at Federal Regulation of Formaldehyde: A Departure from Reasoned Decisionmaking," *Harvard Environmental Law Review*, vol. 7, 1983, pp. 324–46.

17. NRDC v. Ruckelshaus, U.S. District Court for the District of Columbia, no. 83–2039, 1984.

18. *Federal Register*, vol. 48, no. 224, Nov. 18, 1983, pp. 52507–8.

19. "Report of the Federal Panel on Formaldehyde," p. 165.

20. Sheila Jasanoff, *The Fifth Branch of Government* (Cambridge: Harvard University Press, 1990), pp. 10–24.

21. Ibid., pp. 10-29–10-30.

22. U.S. Environmental Protection Agency, *Preliminary Assessment of Health Risks to Garment Workers and Certain Home Residents from Exposure to Formaldehyde*, Washington, D.C., May 1985, draft.

23. *Federal Register*, vol. 49, no. 227, November 23, 1984, p. 46300.

24. Interviews with several EPA personnel, including Carl Mazza, Richard Hill, Cheryl Siegel-Scott, Elizabeth Margosches, and Frederick DiCarlo, Washington, D.C., September–October 1987.

25. SAB, *Report of the Director of the Science Advisory Board for Fiscal Year 1986*, Environmental Protection Agency, Washington, D.C., 1986, introductory page.

26. Ibid., p. 1.

27. Transcript of the Environmental Health Committee Review of the Draft Risk Assessment Document on Formaldehyde, June 26, 1985, statement by Jack Moore, p. 6.

28. Transcript, statement by Jack Moore, p. 15.

29. Transcript, statement by Rich Hefter, pp. 99–100.

30. Interview with Daniel Menzel (EHC member), Durham, North Carolina, March 30, 1988.

31. Ibid.

32. Interviews with Carl Mazza and Richard Hill at EPA, Washington, D.C., October 1987.

33. Response by the Environmental Health Committee of EPA's Scientific Advisory Board to a List of Questions Submitted by the Office of Pesticides and Toxic Substances Regarding a Draft Document, "Preliminary Assessment of Health Risks to Garment Workers and Certain Home Residents from Exposures to Formaldehyde," October 1, 1985.

34. Interview with Jack Murray, President of the Formaldehyde Institute, Washington, D.C., August 18, 1987.

35. Transcript, pp. 16–65.

36. Transcript, statement by Robert Tardiff, p. 97.

37. Transcript, statement by Daniel Menzel, pp. 94–95.

38. Ibid., p. 89.

39. Transcript, statement by Leslie Stayner, p. 76.

40. Transcript, statement by James Swenberg, p. 116.

41. Transcript, statement by Frederick DiCarlo, p. 137.

42. Transcript, statement by Daniel Menzel, p. 135.

43. Ibid., p. 202.

44. Transcript, statement by Robert Tardiff, p. 151.

45. Transcript, statement by James Swenberg, p. 71.

46. Statement by James Swenberg, Head, Department of Biochemical Toxicology and Pathobiology, CIIT, Research Triangle Park, North Carolina, October 25-26, 1988 (Belmont Workshop).

47. Statement by William Farland, Acting Director, Office of Health and Environmental Assessment, U.S. Environmental Protection Agency, Washington, D.C., October 25-26, 1988 (Belmont Workshop).

48. Transcript, statement by Daniel Menzel, p. 162. The Buck and Starr paper was published in *Fundamental and Applied Toxicology*, vol. 4, 1984, 740–753.

49. The Environmental Health Committee report procedure was gathered from interview with Richard Griesemer, Chairman of EHC, Oak Ridge, Tennessee, April 1, 1988.

50. Environmental Health Committee Review Report of the Draft Document, "Preliminary Assessment of Health Risks to Garment Workers and Certain Home Residents from Exposures to Formaldehyde," Washington, D.C., October 1, 1985.

51. Interview with Daniel Menzel.

52. Letter from Richard Griesemer and Norton Nelson to Jack Moore, July 2, 1985.

53. Interview with Daniel Menzel.

54. Response by EHC to List of Questions, October 1, 1985, p. 4.

55. Ibid.

56. Ibid.

57. Life Systems, Inc., "Expert Review of Pharmacokinetic Data: Formaldehyde," Washington, D.C., January 2, 1986, p. 2–1.

58. Interview with Richard Hill at EPA, Washington, D.C., September 1987.

59. Interviews with EPA staff, including Richard Hill, Carl Mazza, Frederick DiCarlo, Cheryl Siegel-Scott, and Elizabeth Margosches, Washington, D.C., September-October. 1987.

60. Life Systems, Inc., "Expert Review," p. 2–1.

61. Ibid. p. 3–1.

62. Ibid. p. 3–9.

63. Ibid. pp. 3-9–4-1.

64. Frederick DiCarlo, "Memorandum on the Expert Review of CIIT Pharmacokinetic Data on Formaldehyde," EPA, Washington, D.C., undated, p. 2.

65. Ibid. p. 1.

66. Interview with Frederick DiCarlo, EPA, September 1987.

67. Interview with Richard Hill, EPA, September 1987.

68. Interview with Daniel Menzel.

69. Ibid.

70. M. Casanova, T. B. Starr, and H. d'A. Heck, "Comments on the Final Report of the Panel Reviewing the CIIT Pharmacokinetic Data on Formaldehyde," Research Triangle Park, N.C., February, 1986, p. 1.

71. Ibid.

72. Ibid, entire document.

73. Interviews with Robert Neal, CIIT, March 1988 and July 1988.

74. U.S. EPA, *Assessment of Health Risks to Garment Workers and Certain Home Residents from Exposure to Formaldehyde,* Washington, D.C., April 1987.

75. Aaron Blair et al., "Mortality among Industrial Workers Exposed to Formaldehyde," Washington, D.C.: National Cancer Institute, 1986, p. 16.

76. House Committee on Energy and Commerce, Subcommittee on Oversight and Investigations, *Formaldehyde Study,* 99th Cong., 2nd sess. 1986, p. 116, as cited in Jasanoff, *Fifth Branch,* p. 10–32.

77. Jasanoff, *Fifth Branch,* pp. 10-31–10-34.

78. Interviews with EPA personnel, including Carl Mazza, Richard Hill, Frederick DiCarlo, Cheryl Siegel-Scott, and Elizabeth Margosches, Washington, D.C., September-October. 1987.

79. Transcript, statement by Herschel Griffin, p. 85.

80. The current stance of EPA regarding formaldehyde was obtained from an interview with Rich Hefter and George Semeniuk of OPTS, August 1987.

81. Ibid.

82. Interviews with EPA personnel, including Carl Mazza, Richard Hill, Frederick DiCarlo, Cheryl Siegel-Scott, and Elizabeth Margosches, Washington, D.C., September-October 1987.

83. Interview with Jack Murray.

84. Interview with James Gibson, CIIT, Research Triangle Park, North Carolina, March 31, 1988.

85. Interview with Jackie Warren, NRDC, New York, N.Y., February 1988.

86. Interview with Carl Mazza, EPA, October 1987.

87. Interview with Robert Neal, CIIT, Research Triangle Park, NC, July 14, 1988.

88. Statement by Robert Neal, President of CIIT, Research Triangle Park, North Carolina, October 25-26, 1988 (Belmont Workshop).

89. Statement by James Swenberg, Head, Department of Biochemical Toxicology and Pathobiology, CIIT, Research Triangle Park, NC, October 25–26, 1988 (Belmont Workshop).

90. Statement by Robert Neal.

91. Statement by Terry Yosie, Vice President, American Petroleum Institute, Washington, D.C., October 25–26, 1988 (Belmont Workshop).

APPENDIX A: PARTICIPANTS AT JUNE 1985 SAB MEETING FOR REVIEW OF FORMALDEHYDE PRELIMINARY RISK ASSESSMENT DOCUMENT

SAB Members

Dr. Richard Griesemer, Chairman
Director of Biology Division
Oak Ridge National Laboratory

Dr. Herschel Griffin
Professor of Epidemiology
Graduate School of Public Health
San Diego State University

Dr. Jack Hackney
Chief, Environmental Health Laboratories
Rancho Los Amigos Hospital Campus of the University of Southern California

Dr. Nancy Kim
Director, New York Department of Environmental Health
Albany, New York

Dr. Daniel Menzel
Director, Cancer Toxicology and Chemical Carcinogenesis Program
Duke University Medical Center

Dr. D. Warner North
Decision Focus, Inc.
Los Altos, California

Dr. Robert Tardiff
Principal, Environ Corporation
Washington, D.C.

Dr. Bernard Weiss
Professor, Division of Toxicology
School of Medicine
University of Rochester, New York

Dr. Ronald Wyzga
Program Director, Electric Power Research Institute
Palo Alto, California

Consultants

Mr. Leslie Stayner
Epidemiologist
NIOSH

Dr. James Swenberg
Chemical Industry Institute of Toxicology
Research Triangle Park, North Carolina

SAB Staff

Dr. Terry Yosie, Director

Dr. Daniel Byrd, Executive Secretary (EHC)

Office of Pesticides and Toxic Substances

Dr. Mary Argus
Senior Science Advisor

Dr. Frederick DiCarlo
Toxicologist

Mr. Richard Hefter
Project Manager for Formaldehyde

Dr. Elizabeth Margosches
Biostatistician

Dr. Jack Moore
Assistant Administrator

Presentations by Members of the Public

Dr. John Clary
Celanese Corporation

Mr. Kip Howlett
Georgia-Pacific

Dr. Harold Imbus
Formerly of Burlington Industries

Dr. Neil Krivanek
Du Pont Company

Dr. Maureen O'Berg
Du Pont Company

CHAPTER 8

NITRATES IN DRINKING WATER

Alon Rosenthal

The 1986 Safe Drinking Water Act (SDWA) Amendments provide a major role for scientific advice in the promulgation of drinking water standards. Like other statutorily mandated review processes, the law imposes an interdependence between EPA and the Science Advisory Board (SAB). Scientific advice in a regulatory context naturally contains an adversarial component. Yet this dynamic must be weathered by both the regulators and the reviewers and a process of accommodation reached if a fruitful, long-term, working relationship is to be fostered.

Nitrate/nitrite was the first substance that the Drinking Water Subcommittee of the SAB considered after being charged by Congress with the review of what will eventually be well over a hundred drinking water standards. Setting a standard for nitrate/nitrite in drinking water raised in a new context some of the most fundamental and difficult issues and dilemmas that have confronted environmental regulation. Questions about "adequacy of the margin of safety," "protection of sensitive populations," and the extent to which "feasibility" should drive environmental standards lay at the heart of EPA's decision.

In this chapter the evolution of the drinking water standard review process will be presented by contrasting the initial encounter between the SAB and the Office of Drinking Water under the SDWA to the present modus operandi. Beginning with a review of the history of drinking water standards in the United States and an examination of the legislative intent of SAB's role within SDWA, the chapter will then focus on the health effects of nitrate/nitrite in drinking water. A description of EPA's proposed standards will be offered and the response of the SAB detailed and assessed. As the current advisory process for drinking water standard setting differs from the dynamics that characterized the nitrate/nitrite review, an epilogue will briefly describe the changed EPA/SAB interaction. Hence problems that characterized this first utilization of scientific advice under the SDWA will serve as a backdrop to which the present activities of the subcommittee can be compared. This chapter offers insights into the role of scientific advice in regulation of a major and complex drinking water contamination problem. It also suggests procedural and substantive norms that can facilitate a more meaningful process for extended external scientific review within a government department.

DRINKING WATER STANDARDS: A HISTORIC PERSPECTIVE

The first comprehensive regulation of drinking water in the United States came in 1914 in the form of a limit on bacteria content in publicly supplied water. Originally known as the Treasury Standards for Drinking Water,[1] the first regulations came in the wake of a well-publicized typhoid epidemic in 1913 involving 144 Great Lakes seamen.[2] The four major revisions of the statute between 1914 and 1974, while expanding the breadth of the legislation, left its basic form unchanged.

The Safe Drinking Water Act that President Ford signed into law on December 4, 1974,[3] was intended to stress the continuity with the previous legislative efforts to regulate drinking water quality. It therefore took the form of an amendment to the Public Health Service Act.[4] As implementation of the 1974 act was delegated to the fledgling EPA, however, it marked a new era for the setting of standards for drinking water quality. National Drinking Water Standards were promulgated on Christmas Eve 1975 and became effective June 14, 1977.[5] The substantial increase in both the number of water sources regulated and the number of substances controlled at the time was remarkable.[6]

RMCLs AND MCLs

The Safe Drinking Water Act, in contrast to exclusively "health-based" statutes such as the Clean Air Act or Resource Conservation and Recovery Act (RCRA), attempts to balance the benefits of providing water at acceptable levels for public health with the costs of treatment when setting concentration levels.[7] Since 1974 the law has distinguished between the ideal concentration or "recommended maximum contaminant level" (RMCL), below which there was no known or anticipated adverse effect on health (allowing for a margin of safety), and a realistically achievable "maximum contaminant level" (MCL).[8] The recently added Section 1412, while changing the name of the first standard to "maximum contaminant level goal" (MCLG) and the second to "national primary drinking water regulations," preserves this conceptual distinction.

The maximum levels specified as national primary drinking water regulations should be as close to the maximum contaminant level goal as is feasible. Unlike the undefined language used in the Occupational Safety and Health Act,[9] the term feasible is here clarified to mean "with the use of the best technology, treatment techniques and other means which the Administrator finds, after examination for efficacy under field conditions and not solely under laboratory conditions, are available (taking cost into consideration)."[10]

A good example of the flexibility that the act allows can be seen in the regulations proposed for control of volatile synthetic organic compounds. The MCLG for benzene is zero, consistent with a nonthreshold dose-response model for this substance's carcinogenic effects. Risk-management considerations led to the selection of 5 µg/l as an acceptable level for the primary regulation. More potent vinyl chloride, on the other hand, has a 2 µg/l primary value, much closer to its zero MCLG. In other words, the greater the magnitude of the risk that a substance poses, the more weight it carries in balancing such factors as treatment feasibility and the chemical's utility to the public.[11]

THE SAFE DRINKING WATER ACT AND THE SAB

The passage in June 1986 of the Safe Drinking Water Act Amendments[12] is widely recognized as a congressional attempt to revitalize a statute that was not satisfactorily fulfilling its legislative mission.[13] The legislation approached the SDWA's inadequacies from many different angles. Its provisions included groundwater protection programs,[14] upgraded enforcement requirements,[15] and even an outright ban on the use of lead pipes, solder, and flux for plumbing and public water systems.[16]

The most immediate and dramatic reform, however, required EPA to substantially increase the pace and number of new drinking water standards. The act specified a schedule for promulgation of regulations for eighty-three substances during the three years after passage. Following this initial promulgation, each year the administrator would be required to issue standards for an additional twenty-five substances. There seems to be agreement among SAB and EPA personnel that the congressional schedule was ill considered, made without adequate regard for the technical resources necessary to ensure the scientific quality of the standards.[17]

These regulations establish the maximum contaminant level allowed in drinking water supplies and are supported by scientific criteria documents. With an estimated 20,000 or more public water suppliers in the United States subject to the new standards, both the economic and health ramifications of this process are enormous. A central actor in this ambitious enterprise is the SAB, which Congress wove into the drinking water standard-promulgation process.

The original statement in the law regarding scientific review required the National Academy of Sciences to provide recommendations to EPA about proposed standards and support documents. This provision was replaced in the 1986 amendments with a more explicit role for the SAB. Specifically, Section 1412 (11)(e) requires that the administrator request comments from the SAB prior to proposal of a maximum contaminant level goal and national primary drinking water regulation. Accordingly, "The Board shall respond, as it deems appropriate within the time period applicable for promulgation of the national primary drinking water standard concerned."

LEGISLATIVE INTENT OF SAB'S ROLE IN SDWA

The ostensible explanation for the replacement of the National Academy of Sciences' authority to review standards by that of the SAB is chronological. While the SDWA was enacted in 1972, the Science Advisory Board was a product of the Environmental Research and Development, Demonstration Authorization Act of 1978.[18] It appears, however, that the unsatisfactory relationship between EPA and NAS may have also contributed to this change. Senator David Durenberger's diplomatic explanation while presenting the Conference Committee report to the Senate may have understated the problems:[19] "The National Academy of Sciences never took well to its assignment to quickly develop explicit recommendations for standards on specific contaminants."[20]

Congress had authorized NAS to receive information from EPA and to advise the agency about drinking water standards. While receiving the required materials and funds for this role, NAS declined to comment on the proposed permissible levels. It perceived standard setting as a fundamentally regulatory function, and as a scientific body it was uncomfortable reducing complex toxicological information to a single number upon which EPA would base its standards.[21] Consequently, NAS reviews were limited to assessments of the quality of the data upon which standards were based. The continuing publication of *Drinking Water and Health,* a summary of the present literature and state-of-the-art technology, was seen as a more appropriate general contribution to the scientific integrity of drinking water policy. The ramifications of this institutional mismatch were profound.

While EPA's Office of Drinking Water was willing to promulgate MCLs without an authoritative number from NAS, agency lawyers were uncomfortable doing so. The attorneys feared that the absence of direct NAS input made EPA vulnerable to legal attacks on procedural grounds. The legal vulnerability, according to one view, left the agency's Drinking Water Program during the 1970s and 1980s "in a state of paralysis."[22] Other explanations for EPA's feeble promulgation record have been advanced. "Terror" of the regulated community, insufficient agency resources and commitment, and a "water supply" rather than a "public health" mentality help explain why only six drinking water standards were promulgated over the first fourteen years of the SDWA.[23]

Congress was dissatisfied with the situation, but was "smart enough to deal with the problem without calling NAS to task."[24] Thus, while the 1986 amendments called for a continued advisory relationship between NAS and EPA,[25] the academy's operational responsibilities and authority were transferred to SAB. Publicly, EPA representatives referred to the change as simply acknowledging an operational reality, calling the previous requirement "out of date."[26]

The House and the Senate bills differed in their vision of SAB's role within the regulatory process. These minor discrepancies were not without their attendant nuances. The EPA staff overseeing SAB activity participated in the discussion on Capitol Hill. As they did not want the obligation of reviewing each and every substance, they lobbied Senator Durenberger to simply offer them the opportunity to review a substance.[27] Hence the Senate bill compelled the administrator to provide SAB an opportunity to comment prior to proposal or during the public comment period. The House amendment required the administrator to request comments from the SAB prior to proposal of a MCLG and national primary drinking water regulations.[28]

Ultimately the House version was approved. It also required that SAB respond as it deems appropriate within the time period applicable for promulgation of the specific national primary drinking water standard. Attempts by industry organizations such as the National Association of Water Companies and the American Water Works Association to further enhance SAB's role in the promulgation process were unsuccessful.[29]

Presumably the requirement of a timely SAB response was largely due to congressional concern about lags within the standard-setting process; the act stipulated that the review process not be used to delay the promulgation of the final standard. Congressional counsel Jimmy Powell explained that the Conference Committee was aware that the SAB had previously been the source of major delays in the implementation of other environmental statutes: "Important decisions in the area of air toxics were delayed simply because SAB had run out of travel money."[30] Insofar as the amendments were designed to accelerate the process and prevent such delays, "the House language succeeded in accomplishing this more directly [than the Senate]."

NITRATE/NITRITE HEALTH EFFECTS

Nitrate is a pollutant that can be traced to both point- and nonpoint-source pollution. Among the major point sources discharging nitrogen into surface waters are municipal and industrial wastewaters, septic tanks, and feedlots. Fertilizer runoff from farms and lawns, animal wastes, and leachate from waste disposal, along with atmospheric fallout of nitrogen oxides emitted by mobile sources and power plants, constitute the major background or nonpoint sources of the pollutant.[31] Since nitrate is one of the two major nutrients responsible for eutrophication of surface waters, concern about nitrate concentration often focuses on environmental and ecological degradation. High levels of nitrates in drinking waters, however, can pose a serious hazard for human health.

Unlike many of the recent substances regulated under the SDWA, nitrate/nitrite is considered a "threshold pollutant," meaning that sufficiently low levels of human exposure to the substance do not cause adverse health effects. Nitrates (NO_3) are most toxic under conditions in which they are reduced to nitrites. The nitrite ion (NO_2) is far more reactive than the nitrate or ammonium ions from which it is formed. The reduction of nitrate to nitrite by microorganisms occurs in a wide variety of systems (e.g., soil, water, sewage) and under certain circumstances in the digestive tract and saliva.

The low acidity in the stomachs of infants (and some adults) allows for colonization of bacteria that can reduce ingested nitrate to nitrite before it reaches the duodenum. When nitrite is absorbed in blood, it can oxidize hemoglobin to methemoglobin, impairing its ability to transport oxygen. When methemoglobin levels in humans rise far beyond background levels of 1 to 2 percent, asphyxia can occur.[32] Infants are the population group at greatest risk of developing methemoglobinemia from high nitrate levels in water. There are several reasons for this. The high pH level in the stomach of infants sometimes allows for a much greater percentage of bacterial transformation from nitrate to nitrite. Furthermore, 60 to 80 percent of fetal hemoglobin is ineffective in the immature form of the molecule, and the hemoglobin that is present is far more readily oxidized than in adults. Moreover, at birth, infants have a deficiency in the level of nicotinamide adenine dinucleotide (NADH), which reduces methemoglobin and returns the blood to normal hemoglobin levels. Finally, infants consume more than five times more water relative to their weight than do the rest of the population.

Animal studies are of little help in determining the levels at which nitrite begins to cause this specific health effect. The rat, for example, is a poor subject for experimentation since rats can reduce methemoglobin much more quickly than humans. Excessive reliance on rat studies would severely understate the health hazard posed by NO_2 to sensitive human populations.[33]

Despite extensive research, no serious chronic effects have been linked to nitrate/nitrite ingestion. Nitrate and nitrite acting alone have not been shown to have carcinogenic effects.[34] Nitrites, however, might give rise to oncogenic chemicals that can cause cancer. Specifically, there is some evidence that suggests that high dosages of nitrite might react with exogenous and endogenous chemicals to form N-nitroso compounds.[35] Numerous scientific studies have confirmed the belief that nitrosamines (a class of amines formed by actions of nitric acids) are carcinogenic.[36]

Considerable attention has been devoted to risk assessments of nitrosamines because of their prevalence in the curing of meats. While tumors were purportedly induced in a number of bioassays involving nitrites, evidence has been marred by reinterpretation of the tissue slides and presence of additional nitrosable foods in the experiment animals' diet.[37] Epidemiological studies have also proven negative or nonconclusive, largely because of the enormous number of potentially confounding factors.[38] Thus, in the absence of any definitive link between nitrates and cancer in humans, nitrate/nitrite are generally classified as noncarcinogens.

THE NITRATE/NITRITE STANDARD AND CRITERIA DOCUMENT

A standard for nitrate was first promulgated with the twelve new criteria authorized in the 1962 Public Health Service (PHS) Drinking Water Standards.

The maximum concentration for total nitrate was set at 45 mg/l,[39] which corresponds to both of the present standards (MCL and MCLG) of 10 mg/l, which only measure the nitrogen component of nitrate. (Though it is confusing to do so, nitrate concentrations in water continue to be measured both as nitrate and as nitrogen.) The Safe Drinking Water Act had adopted the previously existing PHS standards only as "interim standards" and called for EPA to review these levels in light of the wealth of new scientific information and understanding by 1986.[40] EPA proposed to do this in five phases, beginning with volatile organic compounds and then going on to the inorganics group in which nitrate/nitrite was classified.[41] Thus in the early 1980s the review of the interim standard for nitrate began.

Under Section 553 of Title 5 of the SDWA, a criteria document is to be prepared that supports any proposed standard. In the case of nitrate/nitrite, EPA contracted preparation of the criteria document to ICAIR, Life Systems, Inc., a consulting firm in Cleveland, Ohio. Ruth Shearer, a toxicologist who previously had conducted research on the subject, wrote the initial draft. In October 1985 a final draft of the criteria document was printed, with a summary published in the *Federal Register* soon thereafter.[42]

A criteria document is divided into six sections: physical and chemical properties, toxicokinetics, human exposure, health effects in animals and humans, mechanisms of toxicity, and quantification of toxicological effects. The quantification section translates the previous ones into the actual standard proposed, setting forth agency rationale for the concentration limits selected. The quantification section opens with a presentation of the relevant formulas by which EPA usually calculates the "acceptable daily intake" (ADI) and "adjusted acceptable daily intake" (AADI), which correspond to the standards that are ultimately set.[43] Both of these measures are based on the no or lowest observable adverse effect level (NOAEL/LOAEL), with built-in safety factors that reduce them by 10, 100, or 1,000, depending upon the quality of available data.[44] For example, when only animal data are available and they are limited or incomplete, the agency reduces permissible levels below the LOAEL by a factor of 1,000. Accordingly, even when good acute or chronic human exposure data are available and supported by toxicity data in other species, the criteria document states EPA's policy of reducing permissible levels tenfold.

Following this introduction, a synopsis of the most useful studies that might provide the values to put in the equation are reassessed. In the case of nitrate, the investigation that was cited as providing the most compelling evidence and that consequently served as the sole basis for EPA's decision dated back to 1951—the so-called Walton study.[45] The survey by Walton was misrepresented in the original draft as an epidemiological study when in fact it more closely resembled a review of the current literature and information regarding methemoglobinemia.

In 1950 the Committee on Water Supply at the American Public Health Association (APHA) circulated a questionnaire to the forty-eight state Departments of Public Health requesting information on infant mortality associated with drinking water levels. Walton used responses from this survey in qualifying his results, admitting that he did not know over what time frame the statistics were accrued, and concurred with an article stating that methemoglobinemia often goes unreported.[46] The absence of any critique of the methodologies involved in water sampling or operational definitions for disease underscores the limitations in the study. The unavailability of the original Walton article left most of the participants in the standard-setting process unfamiliar with the flaws in the study and undoubtedly contributed to excessive reliance on its results. A summary of these results was given in Chapter 8 of the criteria document, "Quantification of Toxicological Effects":

The study by Walton (1951) on reported cases of methemoglobinemia in infants (diagnosed by observed cyanosis), has been selected to serve as the basis for the AADI. This survey of nearly 300 cases of methemoglobinemia in infants in the United States found no cases associated with water containing less than 10 mg/l of nitrate-nitrogen. This study was selected in preference to other studies of infants because it surveyed a large population and it was not complicated by associated cases of diarrhea.[47]

The document had previously noted that this level was supported by three additional studies.[48]

The document had no difficulty in setting a 111-mg/l ADI for a 70-kg adult based on a 1981 epidemiological study.[49] The RMCL, however, was to be based on the sensitive populations. In contradiction of the aforementioned policy regarding safety factors, the ultimate standard was set without one:

The AADI for nitrate-nitrogen is 10 mg/l. This value is derived by direct use of the NOAEL of 10 mg/l observed by Walton in a study of methemoglobinemia in over 300 human infants exposed to nitrate in drinking water. Since the study involved the most sensitive subgroup of the human population (i.e. infants), no uncertainty factor has been employed in deriving the AADI from the NOAEL.[50]

A more conventional method was used to derive the ultimate RMCL for nitrite. As no study existed suggesting a definitive NOAEL for nitrite, an uncertainty factor of ten was selected and plugged into the Walton-based equation as a denominator, making for a 1-mg/l level.[51] Justification for this value was brief: "This factor has been selected because the NOAEL employed is for nitrate, while nitrite is the toxic species and there is uncertainty regarding the rate and extent of nitrate transformation to nitrite."[52] The document went on to cite a 1972 National Academy of Sciences report that recommended this level for nitrite,[53] as did a previous inquiry undertaken by EPA. The latter points out that waters with significant nitrite concentration usually would be heavily polluted and probably bacteriologically unacceptable.[54]

The section also rejected the evidence of carcinogenicity as "inconclusive." For the fourth time in the document, the results of studies that pointed to low cancer levels among populations with a high intake of fresh fruits and vegetables (which are thought to inhibit nitrosation in animals) were mentioned. Given the many possible explanations for the association, a causal link was ruled excessively tenuous.

PUBLIC COMMENTS ON THE CRITERIA DOCUMENT AND PROPOSED STANDARD

On November 13, 1985, a summary of the document's Section 8 was published as a proposed rule in the *Federal Register*.[55] It asserted that the World Health Organization (WHO) guidelines were identical to the proposed RMCL and that the International Agency for Research on Cancer (IARC) had not classified nitrate/nitrite for potential carcinogenicity. Among the questions raised by EPA for comment were whether the carcinogenic potential of nitrosable compounds should influence the proposed RMCL, and what form the uncertainty factor applied in the calculation of the AADIs should take.

A total of fourteen individuals and organizations provided comments in response to the MCLG (RMCL) proposal. Arguments were put forth attacking the proposal both as too stringent and as too lenient. Municipalities generally looked with disfavor on the proposed standard, viewing it as unnecessary and imposing a heavy financial burden on their already-overextended resources. In its response EPA appropriately did not consider the economic costs of compliance with the MCLG, which, as mentioned, is to be set on the basis of health considerations alone.

The director of public works in Ceres, California, for instance, admitted that most of his city's wells were above the 10-mg/l standard but below the 45-mg/l state standard. (This constitutes an excellent example of the confusion caused by the seemingly different but actually identical standards.) The proposed standard was assailed as forcing the purchase, installation, and upkeep of an activated charcoal filter system, which is both energy intensive and subject to malfunctions that would then create a "serious health threat." Interestingly, the Metropolitan Water District of Southern California, the nation's largest supplier of water, had no objection to the standard. It did, however, acknowledge the difficulty that facilities using chloramine disinfection would have in complying with the standard.

The water utility group of Edison Electric Institute admonished the agency and its logic, asserting that the proposed standard addressed neither the extent of exposure to nitrite nor any projected changes in circumstances. Similarly, the American Water Works Association expressed its view that sufficient occurrence data did not exist to warrant a RMCL for nitrite.

There were several commenters, mainly from state or local public health organizations, who felt that the proposed level was too high. Their basic objection involved the absence of a margin of safety against methemoglobinemia. The agency response reasserted the dearth of any cases of methemoglobinemia at the proposed levels and its position that rejected nitrate or nitrite's oncogenic potential.

A representative of the New York State Legislation on Water Resource Needs of Long Island reproved the standard from a statistical vantage point regarding probable exposure. The inadequacy of the 10-mg/l standard was assailed on the basis of the variability in nitrate concentration over time and space within an aquifer. The average nitrate concentrations should be far enough below the 10-mg/l level to include probabilistic variations in concentrations.[56]

THE DRINKING WATER SUBCOMMITTEE'S PUBLIC MEETING

On August 21, 1986, the Drinking Water Subcommittee met at EPA headquarters in Washington, D.C. The proposed agenda included a review of the draft drinking water criteria document for nitrate/nitrite and a general discussion of the role of the SAB in light of the recent amendments to the Safe Drinking Water Act. While the nitrate document had been prepared under the now-amended provision regarding validation of interim standards, the 1986 amendments also had required the setting of a standard for the substance. Thus the Office of Drinking Water (ODW) decided that the SAB review would be legitimate under the revised terms of the act.[57]

Committee members had received the Criteria Document some time prior to the meeting, and the document had apparently already undergone modest revisions. They did not receive copies of the public comments, although they had occasion to mention their general interest in seeing them.

ADEQUATE MARGIN OF SAFETY

After the initial introductions had been made, Ruth Shearer opened the discussion with a statement that startled some of the participants of the meeting.[58] According to Shearer, the original draft had incorporated a safety factor of ten into the nitrate standard. Since her completion of the draft, the chapter concerning quantification of toxicological effects had been altered, resulting in the dropping of the "safety factor." Shearer's position was that although technically representing the agency in front of the SAB, she disagreed with the standard that was actually being proposed. Rather, the nitrate drinking water standard should include the safety factor and require a level of 1 mg/liter.[59] She explained:

I disagree with dropping the uncertainty factor because . . .the studies selected as the basis of the standard for infants and for the older population . . . were based on healthy people. [Whereas] the sensitive population include the infant, the 4 kilogram stage children with diarrhea, and children that are not supplemented with vitamin C. Sensitive populations are also those with certain enzyme deficiencies, those with cancer, those that are pregnant, those that are taking certain drugs that can interact with nitrate metabolism.[60]

Significantly, Shearer considered her position as consonant with standard EPA policy guidelines, where a safety factor of 10 is used in setting health-based standards based on human data and a factor of 100 is used tor standards based on animal data. She suspected that the proposed standard was motivated by considerations of feasibility rather than actual safety, which were illegitimate given the statutory directives for RMCL safety levels.

Shearer, in a memo to Life Systems, later explained that she had been surprised to find that her name was on the agenda for an opening statement. "Lacking time to develop a tactful approach, I simply told them how shocked I was by the changes made since I last saw the criteria document in January of 1984." She also took umbrage at the manner in which the standard was proposed. "It was frustrating to see how the results of my work were used (and unused), but interesting to see how the new data could be manipulated (by EPA 'upper management' according to Dr. Ohanian) to preserve the status quo in standards."[61]

Her position, and the vehemence with which she stated it, came as a surprise to representatives of EPA. In retrospect, they expressed no resentment toward her presentation, nor did they see an attack by the document's author on ODW's revisions as inappropriate. During a recent review of the polychlorinated biphenyl (PCB) standard, they faced the identical situation when a consultant took issue with the agency's ultimate standard. EPA's justification for this practice is perceived as self-evident: "Remember that the Criteria Documents are fundamentally our property. We own them and we can change them if we need to."[62] Shearer's role in attending the subcommittee meeting was to explain the organization and presentation of the data in the criteria document. Having done so, she was informed by Ken Bailey that she would not be needed for the continuation of the SAB discussion on the following morning.[63]

Bailey had recently been brought on board at the Health Effects Branch of the Office of Drinking Water. He subsequently took charge of the criteria documents for many inorganic compounds, including nitrate, and at the meeting spoke on behalf of the agency. There were three main points that he emphasized in defense of the proposed 10-mg/l standard. As no meaningful studies regarding the long-term effects of nitrate exposure existed, there were insufficient data to establish a typical ADI. The history of the existing standard showed that it was effective in preventing infant mortality. (Ironically, after almost thirty years without a single reported death in the United States associated with high levels of nitrate in drinking water, a few months prior to the meeting, an infant died as a result of nitrate concentrations that reached 150 mg/l in a rural, unregulated South Dakota well.)[64] His main point, however, stressed the fact that only 1 percent of adult exposure to nitrate was from drinking water, with the rest coming from food, particularly vegetables.

Robert Tardiff, who was presiding at the time as chairman of the subcommittee, seized on Bailey's response, suggesting that it was inappropriate for EPA to limit its consideration of health effects solely to mortality. Bailey rejoined by stating that this was the focus because it was the best-recognized problem. Unfortunately, there were simply no data that associated morbidity or what he called "quality of life factors" with nitrate levels. Returning to his initial argument regarding exposure, he explained: "You know, if we set it at zero it will make no earthly difference in terms of your nitrate consumption. Although nitrate for the infant is a serious problem, drinking has very little to do with it simply because of how little drinking water contributes to your ingestion."[65]

Tardiff countered by implicitly arguing that adult exposure was irrelevant. Infants were the population for whom the standards should be designed. Ed Ohanian, director of the Health Effects Branch, chose to field the question. His position was based on the distinction between the application of human and animal data. In the present case solid epidemiological evidence existed and pointed to a clear concentration beneath which no health effects were observed. Setting a level far below this on the basis of "Best Scientific Judgement" precluded the need for an additional safety factor of ten. Had the standard been based on animal data, there would have been a need to build in an additional margin of safety.

The subcommittee did not delve into a serious discussion of the caliber of the Walton study until it returned to the document on the following day. Only then were questions raised as to its validity and accuracy. The notion that it was simply a literature review was raised, yet nobody at the meeting could speak about the study with authority, most admitting that they had not read it.[66] Interviews with SAB members suggested that as the documents are generally read close to the time of meetings, it is exceedingly rare for individual members to contact EPA and request actual primary literature. Under ideal circumstances subcommittee members will receive the documents three to four weeks prior to the review. The documents are sent out only after having been approved by the agency. Typically, this delays the process somewhat, leaving the members with even less time to prepare themselves for the bimonthly meetings.[67] Moreover, the volume of paper that members receive is already substantial. In cases where the scientific merits of a particular piece of experimentation are the subject of controversy, the subcommittee has the authority to request, and has on occasion requested, the text of the study in question.[68] In the case of nitrate it did not do so.

There was a suggestion that the criteria document include a warning for sensitive populations. Tardiff suggested that the agency promulgate advice to physicians and to new mothers on the potential hazards associated with water with a high nitrate standard.[69] Other committee members felt that such an approach's effectiveness would be limited, comparing the situation to asking pregnant women not to take drugs—"the problem is that the people who take drugs become pregnant."[70]

THE SCOPE OF SAB REVIEW

Herschel Griffin suggested that this line of discussion was crossing the line between risk assessment and risk management. Griffin's view of SAB's role presumably was to remain firmly within the assessment side of the review spectrum. Whether or not EPA should prohibit concentrations of nitrates in water that could conceivably endanger very small sensitive populations or simply warn them was ultimately a decision for the risk managers.[71] At a later point in the discussion, this distinction was honed by EPA representatives who suggested that the statute itself made the separation between risk management and assessment. The setting of the RMCL involved strict risk assessment, while the MCL decision, which incorporated economic judgments, was primarily an exercise in risk management.

These issues sparked references to the perennial difference of opinion between the Air Program and the Environmental Health Committee regarding the scope of the SAB review. The Drinking Water Subcommittee felt that among the material covered in the nitrate/nitrite document, there should be information regarding the magnitude of human exposure. For example, there was considerable speculation surrounding the possibility of transport and transmission of nitrate through maternal milk.[72] This would help the committee in targeting the critical effects and the critical ages of the populations. Indeed, at the time of the meeting EPA was developing an exposure assessment for nitrate/nitrite, but it was not yet available.[73] Griffin summarized the drawbacks of a more narrow review process:

The Executive Committee could quite rationally say, well, we'll have the Drinking Water Committee do the health part of it and we'll have the Environmental Engineering part do the treatment technology and we'll have a Committee on Exposure Assessment do that part of it. And we'd then get three documents and sooner or later they would all have to be put together. And I look back on our dioxin experience, I wanted to avoid that.[74]

In contrast, the Air Program's position was that review should be limited solely to toxicological information. Similarly, the subcommittee voiced its frustrations that after review of the document, where many areas of research were shown as inadequate, there was no forum or channel through which they might have input in these critical areas. For example, because the main epidemiological study for nitrate had been conducted in 1951, the members seemed to feel a need to verify the results under conditions of modern diets and water concentrations.

ODW representatives, on the other hand, appeared to support the view that the subcommittee limit its review to the criteria document section on the quantification of toxic effects.[75] Terry Yosie stressed that although the statute limited the SAB's role to the technical basis of standards, he would like to broaden the board's focus beyond toxicology.[76] Issues like treatment technology, research initiatives, and general policy issues should be included in an expanding relationship between the ODW and the SAB.

RESOLVING THE MARGIN OF SAFETY ISSUE

By the end of the morning session the issue of an adequate margin of safety was clearly very far from being resolved. Many committee members voiced concern over the absence of a safety factor, and Tardiff in his summation left it as a "topic for consideration."[77]

On the following day, despite requests by EPA representatives, Griffin, sitting as chair, resisted submission of the question of the adequacy of the margin of safety to a "straw vote." Avoiding a direct confrontation with the agency, he relied on the scientific propriety of the subcommittee: "I really recoil against voting—this is science, it's not democracy."[78] Rather, he felt that the record showed already that there was a general feeling of discomfort with setting a standard so close to the NOAEL value and that there was no point in specifying which margin of safety would be appropriate. Yet when the committee was asked to endorse a recommendation that expressed apprehension that subsets of the population might not be protected by the standard, there was a basic unwillingness to render any decision due to the insufficiency of the evidence.[79] Averting an apparent impasse, Nancy Kim suggested a possible wording for an alternative recommendation that was endorsed unanimously:

Based on the data in the nitrate document, there appears to be little margin of safety in the 10 milligrams per liter level for nitrates for at risk infants. The studies on which the standard is based should be reviewed critically and a rationale provided for the selection of a final number with an estimate of the margin of safety.[80]

REVISING THE CRITERIA DOCUMENT

The overall feeling of the committee was that the criteria document was of extremely poor quality. The point was melodramatically stated by one subcommittee member:

My motivation and enthusiasm is really dampened by trying to react to a document such as this because there are such fundamental organizational even conceptual kinds of things which are out of order that they distract me from getting at the hard facts. . . . If every Criteria Document that we were to react to were at this level, I think that the motivation of the Committee will be destroyed and we won't do our jobs.[81]

Since outright rejection of a document is not perceived by members as a realistic option, a litany of criticism was leveled at the draft. The subcommittee made a number of editorial comments, often correcting the way a study was presented or criticizing the value of a certain study.[82] The time lag between the preparation of the document and the SAB review was cited, and the document was criticized as out of date.[83] Similarly, the excessive length and the organization of the document were criticized by a number of members.[84] The general view of the subcommittee was that the criteria document appeared to resemble a "review of the literature," and like other ones it was inadequate in this capacity, with the sheer quantity of material ultimately only creating confusion.

Ironically, there was a paucity of information about the central studies upon which the standard was based. The need for better documentation of the research methodologies and results of these experiments was deemed important. This would allow the reader to evaluate agency rationale and judgment without having to go back and find the original paper. For example, the use of tables or appendices to contrast results from different studies was recommended in order both to shorten the document and to allow for greater concentration on the key studies. EPA representatives agreed that this form was probably more comprehensible but felt it largely to be already in use.[85]

The committee was perhaps most unified in its concern regarding the absence of an explanatory exposition at the opening of the criteria document. A new member of the committee, Ruth Shecther, expressed her confusion regarding the audience for whom the document was intended.[86] Due to similar complaints, it was suggested that the document begin with a statement that clarified its purpose and function.[87] This would help to signal to reviewers the extent of detail that should be demanded in describing the sundry studies cited and would serve "as a template to gauge the conclusions."[88]

The structural problem was not seen by the SAB as unique to the nitrate/ nitrite document but as one that had plagued previous ODW criteria documents. Although the inclusion of an introductory paragraph met with little opposition by EPA representatives, suggestions to create a standardized, more succinct and understandable format were not warmly received.[89] Over fifty documents were already in the advanced stage of preparation, and EPA saw little benefit in changing the format of the existing drafts at this stage. During the course of the sessions, this issue caused the most friction between members of the subcommittee and EPA personnel. Subcommittee Chairman Carlson saw this as one of the few fundamental issues on which there has not been accommodation by EPA and in retrospect acknowledged that SAB "lost this one."

The implications of these editorial comments were summed up by Griffin on the second day of the meeting:

The environmental health committee has been castigated for delaying the Health Assessment Documents and we have told them over and over again—If the document came in good the first time, it would go right through. This is not laying the blame. It is just a fact. Perhaps there can be mechanisms developed to make prior staff preparation more relevant and appropriate.[90]

THE ISSUE OF CARCINOGENESIS

The subcommittee did not have substantive criticism of the conclusions that the document reached vis-à-vis the cancer-causing properties of nitrates in drinking water. If anything, the members disapproved of the tendency to overemphasize carcinogenicity data.[91] However, it was felt that there should be more open admission of the limits of knowledge. As Daniel Menzel expressed it, the public wants to hear that EPA is concerned about the problem but does not need to hear that a definitive conclusion has been reached.

Issues relating to nitrate's possible carcinogenicity should be seen within the context of the overall difficulty of EPA's position under the act. Joseph Cotruvo explained that the SDWA does not give the EPA much room to differentiate between substances once some initial evaluation of carcinogenicity is made. For example, benzene and trichloroethylene (TCE), having been classified as probable carcinogens, demanded a similar regulatory response, even though the evidence on benzene was much more conclusive. Luckily, treatment costs to reduce TCE levels are not prohibitive, and the agency was not pushed into a serious policy dilemma in differentiating between the two chemicals. The SDWA has a conservative bias that mandates erring on the side of overregulation once a substance has been branded as cancer causing. Thus EPA feels that better evidence will be needed to justify classification of nitrate as a carcinogen since such a classification will lead to truly substantial reductions in the compound's presence in drinking water.

DISCUSSION OF THE SAB'S ROLE UNDER THE SDWA

The subcommittee's criticisms of the criteria document were often linked to the drinking water advisory process itself. Several members expressed dissatisfaction with the timing of the review process. The value of conducting the review subsequent to the receipt of and response to public comments was stressed. Their concern was that SAB review not constitute a final "hoop" through which the document must pass before it advanced from a draft to a final stage. The existing practice was disparaged as analogous to a professor scrutinizing graduate students going through orals.[92] At the time of the meeting, it was unclear whether the SAB would review the nitrate/nitrite standard again after EPA had integrated the subcommittee's comments into the criteria document.

Instead, it was suggested that a more interactive process like that undertaken by the International Agency for Research in Cancer be adopted. The IARC convenes a group of professionals representing the various relevant disciplines required for the review for a week. During that time the panel, with the help of an editorial support staff, produces monographs on ten or fifteen compounds.[93] This approach, unlike the system of SAB review, would help to eliminate the regrettable delays between the writing and review of the document. Yet Brubaker reminded the committee that it should beware lest it become generators of documents and not an oversight committee as the law stipulates.[94]

Terry Yosie, director of the SAB, avoided taking a concrete opinion on the issue. While stressing that he was comfortable with the overall time parameters specified by the act (prior to proposal of the MCL and RMCL in the *Federal*

Register), he agreed that there was a need for a clearer statement as to exactly when in the agency's decision-making process the review should be undertaken. Ed Ohanian from ODW clearly did not agree. The system that he envisioned involved different experts writing different subsections, with a single scientific editor putting the parts together in a coherent document.[95]

THE FUNCTION SERVED BY SAB REVIEW

A discussion of the overall purpose of SAB was undertaken during the final afternoon of the two-day session. Philip Enterline explained that he was uncertain as to the role the criteria documents played in justifying the standard, particularly if it was challenged by litigation. By way of explanation, Dan Byrd put forth his view of the SAB review of the document in this context. He saw the courts' scrutiny as primarily limited to procedural questions, assuring that the scientific support was reasonable and not capricious. They had consistently rejected any active critique of the related science. As Byrd perceived the process, SAB served this function. Failure by EPA to utilize the SAB "puts them in the middle of a mess." It was this argument that he put forth to lower-level staff who complained about the delays involved in the SAB process. "My constant counterpoint is no, it doesn't—it really makes things a lot faster. Because it's gonna take you a couple of months now and it's gonna save you a couple of years later of appearing in Court, testifying in depositions and so forth. I call that kind of function a bullet-proof function."[96] The value that the review served "is not just to improve the quality of the document, but also to certify to the public this is reasonable." This function of providing credibility underlay the importance of preserving an image of independence for the committees.

Enterline was still uneasy about the influence that the review would have. Even if after the review of the draft it remained a "rotten document," as long as the agency had subscribed to the correct procedure, the results were ultimately unimportant to the outcome of any litigation.[97] Byrd disagreed with such a cynical perspective. He saw two grounds on which a criteria document could be criticized: (1) when due to no fault of the staff, available data was poor, and (2) when the staff had not done a good job. From his experience, in the first case the agency management would not set the standard, while in the second the staff would have to do the work over.

FOLLOW-UP OF REVIEW: BUREAUCRATIC MIXUPS AND DELAYS

Following the meeting, EPA representatives strongly felt that the SAB had failed to consider some key studies that they were convinced would assuage many of the concerns voiced at the meeting. A package of additional studies discussed in the criteria document that supported the agency position was sent to Executive Secretary Dan Byrd, who in turn was to forward them to subcommittee members. However, soon after the meeting Byrd left EPA to take a position with the Halogenated Solvents Industrial Alliance. While he could not recall receiving the information, a letter in his name, dated October 19, was sent to the subcommittee with five studies enclosed.[98]

The delay between the meeting and the mailing may have limited its effectiveness. SAB members admit that they do not always have time to read the copious amount of material they constantly receive from EPA. In any event the package does not seem to have had much impact on SAB members' opinions. Most members interviewed, in fact, could not recall receiving the materials.

The lack of an institutional memory at EPA and the foul-ups from the frequent transitions in personnel are commonly recognized problems in the agency, as in the public sector in general. Hence this early effort to foster good will and understanding with SAB by the Health Effects Branch of the ODW appears to have fallen victim to the unfortuitous, but not uncommon, changes in EPA midlevel positions.

Another casualty of this upheaval was a letter to the SAB Executive Committee that the Drinking Water Subcommittee decided to draft at the close of the nitrate/ nitrite review session. Its contents were to summarize the subcommittee's views on procedural revisions needed to improve the review process. By the time Richard Cothern replaced Executive Secretary Byrd and began to coordinate the subcommittee's activities, the letter was forgotten and never composed.

THE WRITTEN REVIEW BY EHC OF THE STANDARD

The Environmental Health Committee's final recommendations regarding the Drinking Water Subcommittee were submitted to Lee Thomas on May 11, 1987, seven months later than had been originally projected.[99] As Dan Byrd was no longer an EPA employee, he was not available to draft the letter, as he had agreed to do during the subcommittee meeting. Nine months passed before the new chairman, Gary Carlson, and Robert Tardiff, now vice chairman, could together compose a summary of the meeting.[100] The key comments of subcommittee members were distilled into a two-page letter to Richard Griesemer, chairman of the Environmental Health Committee. The report touched on the following five general areas:

1. The presentation of data in the document was faulted as ''incomplete, and confusing.'' Since the Walton study was central to the conclusions, it merited an expert review by an epidemiologist, a practice that future criteria documents should follow regarding studies crucial to the proposed standards.

2. The technical rationale for developing a margin of safety was questioned. The subcommittee concluded: ''In the Criteria Document for Nitrate and Nitrite, the Agency selects a margin of safety *that excludes for all practical purposes protection of sensitive members of the population,* namely infants with gastrointestinal disease.''[101] The exclusion or inclusion of sensitive subgroups of the population should be decided on

a general policy level before the application of a margin of safety was made. The letter recommended setting a single public health standard for both nitrate and nitrite. Of all the SAB comments, this suggestion is the only one that appears to have had an operational impact on the ultimate proposed standard.

3. The document's limitation of the exposure assessment to oral ingestion of drinking water was claimed to prevent a comprehensive perspective on the extent of water's contribution to total nitrate exposure and disease in various populations.

4. Recommending that nitrosamine formation be addressed in a separate EPA position paper, the subcommittee stressed the need to eventually update the section on carcinogenicity. The letter also provided an example of an area where new information might already require revisions of the document.

5. The subcommittee suggested that the data gap for reproductive and developmental effects receive mention. It also recommended a clearer statement regarding the issues surrounding fetotoxicity.

EPA RESPONSE TO SAB COMMENTS

In May 1987 the Office of Drinking Water published a revised version of the final draft of the drinking water criteria document for nitrate/nitrite. While the review did not affect the actual proposed standards, comparison of the 1987 and the 1985 final drafts reveals attempts to respond to criticism leveled by SAB at the document's form. Most of the changes in the revised document involved trimming much of the toxicological minutiae from the previous draft. Agency personnel acknowledged the value of the committee's editorial comments, particularly suggested deletions of specific irrelevant studies, which were not part of the formal report to the administrator.[102] The only chapter that was substantially altered was the disputed Chapter 8, "Quantification of Toxicological Effects." Surprisingly, no new study was added; the date of the most recent data used was 1983.

The Walton study was described in greater detail and acknowledged to be a "literature survey which is nearly 40 years old" and has "a number of deficiencies which could potentially limit its usefulness as a basis for deriving Health Advisory values." Yet it was maintained as "the most relevant basis for deriving a Ten Day Health Advisory for the 4 kg. infant." Reasons given for EPA's position included the following:

- The data involved actual U.S. populations and were not confounded by differing diets and lifestyles.

- It was a national study including all regions of the United States, and the 10-mg/l level has been supported in subsequent studies.

- As acute diarrhea in infants is relatively common, it could be assumed that some infants with this illness were among those included in the Walton study. "Thus Walton (1951) provides a measure of the sensitivity of infants with acute diarrhea to nitrate in drinking waters."[103]

Privately, EPA officials justified their continued reliance on the Walton study data based on its historic rather than its scientific merits. Bailey explained: "It has historical significance. The real question is, does it work? For years it was the basis for the standard and it had to be mentioned. Also it signifies the fact that we have 25-30 years of experience—real world data." Ohanian added that SAB might not have sufficiently understood the importance of the data from the study or the chart summarizing Walton's results.[104]

The extensive experience with nitrate in drinking water rather than a theoretically correct process was claimed to have driven EPA's final decision. Ohanian explained: "You can't come up with numbers based on impressions. . . . We need to be exact. What is a health problem, a 5% methemoglobin level, 7%, when does the health effect begin?"[105] In an allusion to Shearer's position, Bailey added: "Her conclusion was that the human diet was intrinsically unsafe. But will it be safe if we lower the drinking water standard? She takes some limited information and runs with it."[106]

ODW officials perceived the subcommittee's concerns about the absence of a safety factor as too theoretical, divorced from the regulatory realities.[107] A recent decision by the European Community to retain its 20-mg/l standard for nitrates was cited as validation of agency judgment. In fact, the agency's decision to implement a nitrite standard of 1-mg/l (despite the suggestion forwarded in the subcommittee's summary letter that the standard be the same as that for nitrate) was offered as proof of ODW's willingness to include safety factors when they were germane. The absence of a safety factor was not seen as a departure from sound scientific practice. Rather, an overly zealous adherence to inclusion of safety factors would reduce scientific integrity:

These uncertainty factors are not absolutes. . . . They have shown themselves to be practical and reasonable, but one is not a hostage to those numbers. One uses them as a departure point when one makes judgments on the data you have. The better the data you have, the less you have to rely on safety factors or uncertainty factors. . . . It isn't an automatic system. If that were the case we could do all of this by computers. Not even computers, you could just use calculators.[108]

Thus the justification for the standard was based on scientific/health-based considerations rather than economic ones. On the other hand, the EPA representatives did not hesitate to cite reports that projected additional treatment costs as high as $27 billion to meet a 1-mg/l nitrate standard.[109] To what extent these considerations colored their position from the outset is uncertain.

THE FORM OF AND AUDIENCE FOR SAB RECOMMENDATIONS

Of the numerous suggestions put forth by the SAB during the public meeting, only a few were summarized in the Carlson and Tardiff letter of May 11, 1987.

For the most part, EPA's revisions of the criteria document responded only to these official written comments. This relatively narrow response by the agency to the subcommittee's formal conclusions is the official policy of the Office of Drinking Water in general when dealing with SAB review.[110] It is defended by outsiders as important in order to facilitate open discussion by SAB members. Frequently there are situations where a scientist will have difficulty precisely articulating his position or simply wishes to take a "straw-man" position. To give the oral comments any sort of binding force would inhibit the scientific dialogue. Second, the potential for mischief by attorneys, misconstruing oral comments by SAB members during subsequent litigation, would be almost unlimited.[111]

Other members of SAB disagree with this approach. As Tardiff explained it: "We've never told them, those are the only things that are to be addressed, but rather, those are things we want to make sure you don't fail to address."[112] The lack of an editorial "laundry list" in the official summary was largely motivated by a conscious desire by the subcommittee to avoid a pejorative tone. In Tardiff's opinion this should in no way have distracted the agency from the less important commentaries that arose in discussion. Nevertheless, he acknowleged that EPA has gotten better at picking up hints from the subcommittee and incorporating smaller "editorial" comments into later criteria document drafts.

The audience to whom the Drinking Water Subcommittee addressed its comments within the criteria document was also unclear. Terry Yosie and Dan Byrd often quarreled as to whether the letters with recommendations from the SAB should be directed at the staff scientist within the agency or at the administrator.[113] While today this question is no closer to being formally resolved than it was during the nitrate/nitrite review, SAB's perception of increased EPA responsiveness to oral comments has to a certain extent made this a moot issue.

THE PROPOSED STANDARD: AN UPDATE

EPA reproposed the MCLG and proposed MCL for nitrate along with 30 additional synthetic organic chemicals on May 22, 1989 (1154 FR 22062). The notice maintained the 10-mg/l MCLG and primary standard of 10 mg/l and 1 mgl for nitrates and nitrites, respectively. The notice also drew new support from a major 1987 independent review affirming the adequacy of the 10-mg/l standard, published by Anna Fan and her colleagues at the California Office of Environmental Health Hazard Assessment at Berkeley.[114] The proposed notice added a 10-mg/l MCLG for total nitrate and nitrite (i.e., the sum of nitrate and nitrite may not exceed 10 mg/l). It would appear that this addition was a direct result of the SAB suggestion.[115]

Each of SAB's written comments was summarized and addressed in the notice, with EPA basically restating its earlier position. For example, in response to the SAB's call for a more extensive analysis of the Walton study the notice explained:

Walton is the historical basis of the current 10 mg/l nitrate standard. While by current standards Walton may be of marginal technical quality, substantial new information since the Walton study was considered in the EPA Nitrate/nitrite Health Criteria Document. This information further supports EPA's conclusion that a 10 mg/l nitrate standard is protective.[116]

Regarding SAB's criticism that the margin of safety used by EPA might not be adequate to protect sensitive members of the population, namely infants with gastrointestinal disease, the agency invoked international support for the standard. The agency admitted that the margin of safety that supported the proposed standard was low. Yet WHO guidelines and Canadian governmental studies, as well as other scientific literature, were referred to as affirming the adequacy of the margin of safety.[117]

SAB concerns about lack of data regarding reproductive and developmental effects were also acknowledged. Yet the agency position held that available data supported EPA's conclusion that at the present standard such effects were not evident. Because the nitrate standard was only part of a larger package of proposed MCLs and MCLGs, the comment period following publication should be longer than usual.[118]

In the meantime, two years after SAB's session on nitrate, EPA's own Office of Program and Policy Evaluation (OPPE) entered the debate. OPPE's interest was sparked by a routine report written by a 1988 summer intern and containing a review of the toxicological basis for the nitrate standard. In the internal report considerable attention was directed at an Australian epidemiology study that suggested a threefold increase in fetal malformations with high nitrate levels in drinking water.[119]

The OPPE, largely unaware until the summer of 1988 of the absence of a safety factor, began to lobby the Office of Drinking Water to promulgate a 1-mg/l standard. ODW, already familiar with the 1984 Australian epidemiological study (as well as several negative ones), has remained unmoved by OPPE's interest.[120] Whether the current internal rift will affect the final standard remains to be seen.

It appears that the ODW position has not wavered since its original revised criteria document in 1985. Since the revised final draft was published in 1987, it has publicly expressed its commitment to a 10 mg/l standard. Thus there has been a considerable period of time during which involved parties can assess EPA's response to SAB's comments. Not surprisingly, the diversity of opinions about the present proposal suggests that many of the controversies surrounding the nitrate standard have not been resolved by the scientific review process.

REACTIONS TO THE PROPOSED STANDARD

The economic impact of a reduction in the MCL varies geographically and colors the opinions of state drinking water officials. North Carolina, with low background levels, would anticipate little additional expense by ratcheting the present standard down to a 5-mg/l level. John Sheets, of the North Carolina Department of Public Health, has claimed that the nitrate MCL is "one of the few that is higher than it should be" and has found EPA's inconsistency baffling:

It looks to me sometimes that they split frog hairs—they make the limit so damn low when the risks are low—they chase goats. And then they turn around in a situation like this and they don't put that margin of safety in. In this case, you're not talking about a low level chronic exposure that may give somebody cancer at the age of 72 when they should have lived until 78. . . . When you start talking about a baby 3-4 months old, it still has fetal hemoglobin.[121]

The American Water Works Association (AWWA) represents a broad interest group that primarily includes drinking water utilities, lab technicians, and utility personnel. While many of its constituents would prefer a higher nitrate standard, EPA's decision to preserve the status quo will not impose new prodigious expenses on most utilities. SAB's discomfort with the margin of safety, however, presents an interesting dilemma for the group. As SAB reviews generally criticize the excessively conservative EPA standards, AWWA has made a policy of deferring to expert judgment by linking its position on health effects to SAB's recommendations. Whether in the interest of consistency it would criticize the proposed 10-mg/l standard poses a currently unresolved dilemma.

AWWA's major frustration with the agency lies in the toxicological orientation of ODW personnel who end up writing regulations that AWWA feels should be designed with an appreciation for the engineering problems. For example, nitrite can be formed in anaerobic environments that are most likely to be found close to the points of delivery in the distribution systems. This raises the question of the boundaries of utility compliance, which are not addressed but may become an important issue with the exceedingly stringent nitrite standard.[122] The Drinking Water Subcommittee has only recently attempted to include members whose expertise involves water supply engineering.

A reduction in the nitrate standard would apparently be disastrous for many public water suppliers. A recent report within the Massachusetts Department of Environmental Quality Engineering described the newest treatment technology for reduction of nitrates and the associated costs. The technology of choice, the rotating biological contractor (RBC), while easy to build and maintain, probably cannot lower concentrations below 10 mg/l. Static-bed up-flow reactors that utilize biological denitrification have also been demonstrated effective in removal of nitrate from water supplies.[123]

Ceres, California, a small community ninety miles south of Sacramento, faces a problem typical of cities in the California Central Valley or at any public supply coming from groundwater sources with high background levels. The ten wells that provide water to the city lie in dispersed locations. Air stripping may be required if the standard is lowered, but there physically is not space on the 60 by 100 foot lots to fit a system and a booster that will restore the water to system pressure. The increased development and continued reliance on septic tanks has led to projections of exceedances in at least half of the wells by 1995. A switch to surface water supply systems would at the very least triple residential water bills. The Department of Public Works distributes bottled water to families with infants when monitoring shows temporary exceedances. Joe Hollstein, director of the department, is aware of the potential for deleterious health effects but is skeptical about EPA's appreciation of the very real engineering problems hundreds of communities like Ceres face.[124]

This gap between the toxicological orientation of ODW and SAB and the engineering realities of water supply policy is the subject of criticism by observers in related fields outside the agency.[125] Thus the real-world problem of compliance and sampling is a crucial factor in the nitrate standard that was not even raised by the subcommittee. For a carcinogenic contaminant, relative risk is not affected by variances in mean concentrations if an assumption of a linear dose-response curve is made. Thus the question of spatial variability of pollutants within aquifers has little importance in the design of control strategies and standard setting. Nitrates are fundamentally different. When a threshold within a sensitive population is driving the standard, deviations from mean concentrations are not only meaningful but can literally distinguish between a safe and an unsafe well.

Keith Powell, director of the Water Resources Institute at Cornell University, has done extensive study and modeling of nitrate concentrations in aquifers. His work suggests that in order to avoid any exceedances of the 10-mg/l standard, mean concentration will have to be very close to zero. The problem as he sees it is one of communication. Toxicologists perceive the standard as a ceiling; engineers see it as an average. These questions, raised during the comment period, were not part of the SAB frame of reference. While not a public health specialist, Powell is concerned about health effects if in fact water supply systems are designed to only have "average concentrations" of 10 mg/l.[126]

SAB'S PERSPECTIVE ON THE EPA'S DECISION

EPA's failure to change the proposed nitrate standards in response to the concerns expressed in the subcommittee's letter was not warmly received by SAB members. As former chairman Robert Tardiff expressed it:

There are two issues: one—ignoring the fact there is virtually no margin of safety, and the other—the way in which it was presented. I think it is a departure from agency policy. I can't think of any other regulation in which the Agency has not in fact taken a more conservative view—that they wanted a greater margin of safety than others might have preferred. To go to virtually none doesn't seem like it complies with the Congressional intent of the SDWA. . . . What no margin of safety means is that there is very little opportunity to take into consideration the variabilities and susceptibilities of the general population. So I think that's kind of unusual.

What's less unusual and from my point of view unfortunate is the way the Agency frequently responds to SAB comments. They feel deeply obligated to provide a description of why they disagree for fear they are going to get sued or publicly chastised—But [there is] also substantial reluctance to make any modifications. It's unclear to me what the underlying reason is for going to the SAB, soliciting their views and always objecting to incorporating those views into the improvement of the document.[127]

Other observers agreed that the nitrate case established a pattern of agency indifference to the Drinking Water Subcommittee's advice: "We would like to see more interaction with them [SAB]. To date [they] have not had a voice in what EPA is doing. It is simply, this is what we are doing SAB, count your blessings, goodbye. We're doing it anyway."[128] While SAB members pay lip service to ODW's authority to disagree with their advice, their resentment at the failure to modify the nitrate standard after SAB's comments is unmistakable.

THE RESPONSIVENESS ISSUE

A variety of explanations have been postulated for the agency's disregard for SAB advice. Some see EPA's response as linked to the institutional arrangements that give rise to criteria documents. In most cases the initial preparation and writing of the documents is left to contractors. According to critics, this leaves EPA personnel inadequately familiar with the properties of the substance being regulated. Dan Byrd, disturbed by the present dynamic, explained:

Ideally the person who does the review should set the standard. Part of what is wrong at EPA is the tendency to contract out too much stuff. The associated problem is that if local officials and people call about an emergency or spill, EPA doesn't have the scientists who can respond because they haven't done the review. The people who are knowledgeable aren't there to field these calls.[129]

EPA vigorously rejects this criticism. From its perspective, extensive "in-house" expertise guarantees that the use of contractors will not render a project manager incapable of making necessary revisions after SAB review. The several doctors of toxicology within the office should have a "better perspective than the person who wrote the document" and are credited with the writing of several complete criteria documents on their own.[130] While acknowledging that a project manager "overseeing preparation of as many as five documents simultaneously may not be as fully familiar with each document as if he'd written each one," ODW does not perceive lack of familiarity with or reluctance to change a document per se to be a problem within the office. In the case of nitrate, the ODW staff certainly felt sufficiently familiar with the scientific issues to depart from the conclusions reached by the contractors preparing the criteria document. Nevertheless, this perception is not shared by many outside observers,[131] as well as SAB members who describe the expertise and involvement of project managers as varying dramatically among individuals.[132]

The primary advantage of the present contracting system is that it allows for utilization of a variety of disciplinary expertise. Thus the salient health effect, theoretically, can determine an appropriate author. For example, evaluating the health effects of a substance like lead requires proficiency in the areas of behavioral and reproductive toxicology rather than the more typical areas of cancer and mutation. If EPA were to rewrite the document, it would be free to employ appropriate expertise accordingly.

Unfortunately, the current bidding system has allowed some consulting firms to take advantage of a project manager's unwillingness to insert subjective judgments in the selection process. Objective reports of firm resources can exaggerate actual capabilities; thus firms will receive sufficient technical scores to secure a passing grade. Although their lower price may win the contract, they may not be able to deliver a document of as high a quality as SAB (or EPA) might like.[133] This has led one SAB member to suggest that SAB be involved in the selection of the contract agent.[134] A more cynical view from Congress perceives EPA reliance on contractors as part of the Reagan administration's larger strategy of keeping bureaucracy small "on paper" while at the same time dismantling the internal scientific research program of EPA.[135]

Another reason why SAB recommendations may not be adequately incorporated into agency documents and policies is the sheer magnitude of the present standard-setting task. There is almost universal agreement that "only a Congressional staffer could have dreamt up the timetable for MCL promulgation." Indeed, Richard Cooper, a seasoned participant in regulatory politics, has observed that tough schedules, like those required under the SDWA, make for "great oversight" and "innumerable opportunities for Congressmen to attack EPA on the nightly news."[136]

As a result, project managers have been described as overwhelmed by the work load: "With the number of documents coming through, they end up just doing a job and not showing any real passion on the subject."[137] Dan Byrd, commenting on the inevitable trade-off of quality for quantity, likened present ODW efforts to an overrun factory: "You know that if you go to the Ford assembly line which is putting out 20 cars a day the best it can and you tell them, we need 80—either you'll get 23 cars with a lot of errors or people will quit. Eventually a new assembly line has to be built."[138] Incorporation of SAB comments involves a substantial time commitment on the part of agency personnel. Thus one ramification of the EPA's current trade-off of quality for quantity may be the underutilization of SAB advice.

A more cynical explanation of the agency's position in the nitrate case perceives the agency's scientific position as reflecting the costs of treatment despite the statutory directive not to do so. Subcommittee member Nancy Kim also serves as director of the New York Department of Environmental Health. Her experience at the state level suggests that EPA's scientific judgment in the present case was ultimately colored by economic considerations.

Having looked at nitrate before, I think that's a lot of what is going on. Probably it isn't feasible. . . . There is a lot of nitrate exposure whether you have 10 or 20 milligrams in drinking water; it's very difficult to remove if it is in the water supply. There is no easy way to treat like there is with VOC's (volatile organic compounds). There's a lot of natural sources. So the risk assessment and the management got a little bit mixed up.[139]

IMPROVED UTILIZATION OF SCIENTIFIC ADVICE IN SETTING DRINKING WATER STANDARDS

An examination of the nitrate/nitrite case suggests a scenario where the SAB is functionally irrelevant within the drinking water standard-setting process. In contrast, there is a consensus among SAB members and observers that the overall trend within the ODW is in the direction of greater responsiveness to SAB's advice, even on the central, substantive issues. As one typical member described the improvement: "It's not glacial in its changes, but it is definitely perceptible and for the better."[140] Among the cases cited where subcommittee comments altered an operational regulatory decision are ethylene chloride,[141] dichloroben-zene, and the use of secondary standards for compounds like xylene.[142] In the remainder of this chapter, the evolution of the SDWA review process will be examined. The focus will be on procedural and institutional modifications that have distinguished subsequent review of drinking water standards from the original examination of nitrate/nitrite.

THE ROLE OF PROCEDURAL CHANGES IN IMPROVING RELATIONS

While EPA still does not agree with the SAB approach in determining the adequacy of a safety margin, the more general criticisms made about SAB's role in the process seem to have had operational effects within the agency. There is widespread agreement on all sides that the subcommittee's review process has gotten more efficient. Long delays between the actual review and the drafting of the recommendations to the EHC have been reduced by having subcommittee meetings follow an intensive two-day format. On Thursdays the agency makes its presentation, and the standard and criteria document are reviewed. On Friday a written report of the conclusions is drafted.

More importantly, the timing of SAB involvement during the standard-setting process has changed. EPA project managers receive input from the SAB as soon as the documents are ready for any sort of external review. When the SDWA Amendments institutionalized SAB review, many criteria documents had been in preparation for some time; the subcommittee could only evaluate a draft in the final stages of its development. Joe Cotruvo, director of the Criteria and Standards Division, is credited by SAB members with changing the procedure for the second and third groups of the eighty-three substances, so that whenever possible, SAB participation begins in the nascent stages of criteria document development. In fact, this procedural change only came when then Executive Director Terry Yosie exercised his statutory authority to ensure that SAB review preceded that of OMB for drinking water standards.[143]

Ironically, there is a faction within the subcommittee that feels that present SAB review is often premature, requiring time-consuming editorial corrections that could just as easily be made by the other committees.[144] Previously, reviewing developed drafts, SAB could only ensure that the "science has not been subverted." Today the subcommittee can increasingly provide early assessments of whether "the authors are headed in the right direction of scientific interpretation."[145]

Responding to the subcommittee's frustration, EPA has begun to routinely include a foreword in each of its drinking water criteria documents, explaining its purpose and providing a synopsis of its conclusions. Subcommittee members and objective observers agree that the quality of the criteria documents is generally getting much better.[146] Many are quick to credit EPA project managers, rather than individual contractors, with the improvement. Generally, contractors are given a very detailed format to follow, which has led to more concise and coherent documents.[147]

Subcommittee members also express general satisfaction at the way in which the agency has come to utilize their expertise. Early resentment at playing a largely editorial role has subsided with their increased participation on the "meatier and more substantive scientific issues."[148] The agency's working procedures with the SAB have become more focused—carefully framing the problems rather than presenting the subcommittee with open-ended questions. Cotruvo has explained:

As opposed to saying "We have a complicated toxicological situation with nitrate; what do you think we ought to do with it?" we will come to them and say "we've done this work and boiled it down to two or three reasonable options and here are the strengths and weaknesses of each of those options. What do you think?"[149]

TOWARD ACCOMMODATION

Both sides agree that the personal relations between the SAB and EPA have improved tremendously since the initial dispute over nitrate, when EPA personnel admitted that the two did not share a common vision of the function of the review process.[150] While the scope of the subcommittee's review goes beyond mere advice on quantification of toxicological effects (as advocated by Ohanian during the nitrate session), certain boundaries have been established. Limiting review to the area of risk assessment, an open issue during the nitrate hearings, has become accepted practice. When issues that fall under the general category of risk management are inevitably raised by members, meetings are steered toward review of the risk management's scientific underpinnings. For example, should agency inferences that are clearly "risk-management considerations" be made under the veil of scientific fact, SAB will dispute them. In retrospect, the criteria document's treatment of the safety factor for nitrate was such an instance.

While SAB is aware of occasional gaps in expertise within the committee, the breadth of issues that come before the subcommittee requires a diversity in the membership that is difficult to meet. Current chairman Gary Carlson has explained that while two engineers have been added recently, they are probably bored during the elaborate discussions of health effects. On the other hand, during recent discussions of best available technology or appropriate analytical methods, the toxicologists have often felt as if they had little to contribute. The loss of specific individual members with broad interdisciplinary expertise has hurt the committee in this regard.[151]

It is unclear whether or not a proposed standard is more or less vulnerable to legal attacks after the SAB process is complete. The ODW sees the process as strengthening EPA's position in court. The agency has gone through "the hoop of review" and technically responded to SAB's comments. In the case of nitrate, with other public comments assailing the standard for being too stringent, SAB's comments are not seen as decisive within the courtroom:[152] "All we have to do is show reasoned information."[153] On the other hand, SAB's criticism would undoubtedly have the effect of shifting the focus of an administrative judge to the disputed issues.[154] The agency will need excellent scientific rationale if it is to successfully reject SAB's advice. In a sense, by expressing its discomfort over a regulatory decision, SAB has the power to reverse the burden of proof. Thus, hypothetically, were the nitrate standard to be challenged, EPA would immediately find itself on the defensive, justifying its indifference to SAB's concern for a greater safety factor.

Although some EPA officials publicly contend that the ODW sought advice from SAB even before it was required to and has never perceived it as an adversary,[155] the first review sessions after the amendments were clearly contentious. The atmosphere in these sessions has changed for the better, as EPA's Bailey has jocularly explained: "We are moving towards detente with the subcommittee. Today it is more of a partnership than what it used to be—which was 'War'. In fact I haven't seen one visit that didn't help us."[156]

THE TIMING PARADOX

It appears that as individual project officers have come to know subcommittee members, telephone calls back and forth and troubleshooting discussions have reduced the formality and the friction in meetings. The increased consultation at the early stages in the preparation of the criteria document between the subcommittee and the ODW raises the fundamental paradox inherent to the review process. Peer review in the late stages of a standard-setting process poses the risk that agency judgment or even the very approach to the problem chosen will meet with unequivocal disapproval by the expert panel. At this stage, revisions may be very costly, requiring duplication of effort and waste of public resources. Even if there is a desire to implement the proposed changes, there may not be adequate time or funds to do so.

On the other hand, SAB involvement at the early stages of a standard-setting process, when the agency position is still uncertain, generally requires a more substantial time commitment. More importantly, it implies that the reviewing body may end up participating in the actual policy decisions. In doing so, reviewers may prematurely come to feel that they have a certain stake in a particular outcome. This can create doubts about their status as independent scientists, detached from a specific approach or consequence associated with a substance's regulation. Many observers feel that there is no solution to this dilemma, but within EPA typically a cycle has evolved where the SAB oscillates back and forth between these two chronological extremes.[157] As Dan Byrd has bluntly explained: "They can't get in bed with the Agency because then they will lose their independence and their value to the Agency. But, over a period of time, if they keep dumping on the Agency scientist, then the scientists stop paying attention to them and won't use their advice. So it waxes and wanes."[158] There is little question that the nitrate review constituted an extreme example of independent effort by the agency and an adversarial review by SAB.

For the most part neither EPA officials nor SAB staff see the trend toward early participation as endangering the subcommittee's independence. Among the safeguards to its independence cited are the rotation of SAB membership, the limited scope of review, and the natural propensity of members to offer criticism, even when there is ostensibly little to criticize. As one SAB member wryly admitted: "I guess the SAB members probably feel a lot of times that if they don't find something wrong, they're not doing their jobs."[159]

THE EMERGENCE OF A SUCCESSFUL SCIENTIFIC
REVIEW STORY

An overview of the two-year history of SAB review of drinking water standard setting reveals an unsettled yet improving process. Certain dilemmas, such as ensuring a membership with specialties in a range of drinking water issues without sacrificing the necessary toxicological expertise, have yet to be reconciled by the subcommittee.

Given the current pace of standard setting, it may be unrealistic to expect SAB advice to be immediately incorporated into agency proposals. Nancy Kim suggests that SAB's most valuable function is rarely seen in the revisions of a specific document under review, but rather in the thought process that is generated in the agency and carried over to the next document. According to this paradigm, EPA's current emphasis on pharmacokinetics is the result of SAB prodding several years ago.[160] The case of the review of the drinking water standard for nitrate/nitrite might best be interpreted from this perspective.

As the first substance reviewed by SAB under the SDWA Amendments, the nitrate standard began a process that would fundamentally change the face of drinking water regulation in America. In retrospect, many factors, ranging from the existence of a complete criteria document to the substance's toxicological nuances, guaranteed that this initial session would be confrontational. Whether or not the agency would have more readily accepted SAB advice if nitrate had been reviewed at a later date, when the review process was more firmly established, is an interesting hypothetical question. It does appear, though, that the frank exchange of opinions on issues surrounding the nitrate review began a process of mutual accommodation. After almost two years of adjustments, personnel rotations, and extremely intensive work, the relationship has blossomed into a fruitful one. EPA officials perceive review as increasingly helpful, and SAB members feel that they are making more meaningful contributions to improving the agency's scientific integrity.

Personnel from the Office of Drinking Water at EPA maintain that even with the SAB review process, they will meet the statutory standard-setting schedule.[161] If they do so, it will to a certain extent be due to a modus vivendi that has evolved between SAB and the agency. Both speak of a learning curve along which the other is progressing. Yet anecdotal remarks reveal an increased understanding on each side of the particular dynamics and dilemmas the other faces. One SAB member likened EPA's position to that of a surgeon who has received contradictory diagnoses about the malignancy of a tumor from two eminent pathologists: "The problem is with tough decisions—experts can vote, but ultimately one person has to decide."[162] While EPA retains its authority, SAB advice is serving two valuable functions. During a period when a frenzied promulgation schedule might undermine the quality of the agency's work, peer review serves to improve the documents supporting drinking water standards. Perhaps more importantly, the scrutiny of the Drinking Water Subcommittee provides "a unifying filter" through which decisions in the standard-setting process must pass. Despite the diversity of actors involved in regulation of drinking water, the importance of SAB review constitutes a rare source of agreement. SAB advice, therefore, serves to enhance these parties' perception of the scientific merits underlying the national drinking water standards.

NOTES

Assistance from the Health Effects Branch in the Office of Drinking Water, U.S. EPA, is gratefully acknowledged.

1. Interstate Quarantine Regulations, Par. 16, Amended June 4, 1914.

2. See Floyd B. Taylor, "Drinking Water Standards: Principles and History, 1914–1976," *Journal of the New England Water Association*, September 1977.

3. P.L. 93–523.

4. P.L. 410.

5. 40 CFR 141.

6. The 1962 Public Health Service (PHS) standards applied only to 700 interstate carrier water supplies used by about 85,000,000 Americans. The EPA standards applied

to 24,000 communities' drinking water supplies, serving approximately 155,000,000 people. Of equal importance was the increase in the number of substances regulated. While the 1962 PHS standards included values for twenty-two substances, thirteen were purely "recommended" and not binding. The new "interim standards" contained many previously unregulated pesticides and involved twenty-one mandatory standards.

7. Kenneth Gray, "Drinking Water Act Amendments Will Tap New Sources of Strength," *National Law Journal*, September 1, 1986.

8. See John Thompson, *Summary of Revisions of the Drinking Water Regulations and Amendments to the Safe Drinking Water Act* (CDM) June 1986.

9. Recent well-publicized decisions by the Supreme Court have begun to flesh out a more complete posture of the toxic policy expressed in Section 6(b)(5) of the Occupational Safety and Health Act (U.S.C.A. 651-55, P. L. 91-596). The clause that has been the subject of controversy regarding the extent of the secretary's discretion when setting a standard for a toxic refers to a standard that "adequately assures to the extent feasible, on the basis of the best available evidence, that no employee will suffer material impairment of health or functional capacity, even if such employee has regular exposure to the hazard dealt with by such a standard for the period of his working life."

10. Safe Drinking Water Act, Sec. 1412(5).

11. The U.S. EPA Health Advisory for Benzene sets excess upper-bound lifetime cancer risks of 10^{-4}, 10^{-5}, and 10^{-6} corresponding to benzene in drinking water at concentrations of 70, 7, and 0.7 μg/l, respectively. The Vinyl Chloride Health Advisory sets excess risks of 10^{-4}, 10^{-5}, and 10^{-6} for vinyl chloride in drinking water at concentrations of 1.5, .15, and .015 μg/l, respectively. See "Benzene, Health Advisory," U.S. EPA, Office of Drinking Water, March 31, 1987, p. 9, and "Vinyl Chloride Health Advisory Draft," U.S. EPA, Office of Drinking Water, March 31, 1987, p. 9.

12. 42 U.S.C.A. 300f et seq.

13. See Gray, "Drinking Water Act Amendments"; presentation of Conference Report by Senator Durenberger, *Congressional Record—Senate*, May 21, 1986, p. S6284.

14. See Safe Drinking Water Act, Secs. 1427 and 1428.

15. Sec. 1423.

16. Sec. 1417.

17. Sec. 1412(3)(A).

18. Interview with Jimmy Powell, Professional Staff, Senate Environment and Public Works Committee, May 16, 1988.

19. Interview with Daniel Menzel, former Chairman of the National Academy of Sciences Drinking Water Committee, February 12, 1988.

20. *Congressional Record—Senate*, May 21, 1986, p. S6286.

21. Interview with Daniel Menzel.

22. Interview with Dan Byrd, Halogenated Solvents Industrial Alliance, January 21, 1988.

23. Interview with Jimmy Powell.

24. Ibid.

25. Senator Durenberger's comments, *Congressional Record—Senate*, p. 56286.

26. Joseph Cotruvo, Director, Criteria and Standards Division, Office of Drinking Water, Comments to the Drinking Water Subcommittee of the Environmental Health Committee, August 21, 1986, p. 123 of official transcript from meeting (reported by Free State Reporting Inc., Washington, D.C.) (hereinafter the first day of the subcommittee

review will be cited *Transcript*, August 21, 1986, and the second day *Transcript*, August 22, 1986).

27. Comments by Terry R. Yosie, Vice President, American Petroleum Institute, at Belmont Workshop on Scientific Advice for Environmental Health Regulation, October 26, 1988.

28. See Amendments to the Safe Drinking Water Act H.R. 1650, March 21, 1985.

29. See statements of Robert A. Gerber, President, National Association of Water Companies, to the House Committee on Health and the Environment, and testimony of the American Water Works Association in *Hearings before the Subcommittee on Health and the Environment*, May 1, 1985, Serial no. 99–28, pp. 213 and 173–74.

30. Comments Terry R. Yosie.

31. U.S. EPA, Office of Water, *Quality Criteria for Water, 1986*, 44–5–86–501, Washington D.C., May 1986, "Nitrate." For a comprehensive discussion of nitrate sources that contribute to high concentrations in surface and groundwater, also see *Drinking Water and Health*, vol. 1, National Academy of Science, Washington, D.C., 1977, and C. J. Downes, "Redux Reactions, Mineral Equilibria, and Groundwater Quality in New Zealand Aquifers, in *Ground Water Quality*, edited by C. H. Ward et al., John Wiley and Sons, New York, 1985, Chapter 7.

32. Shuval, Gruener N. "Health Effects of Nitrates in Water," Cincinnati, Ohio, Health Effects Research Laboratory, 1977, U.S. EPA, EPA 600 18:340–47.

33. Cats' metabolism, on the other hand, goes to the other extreme. Its slower methemoglobin reduction would point to a lower threshold for humans.

34. P. Frazier, "Nitrates, Epidemiological Evidence," in *Interpretation of Negative Epidemiologial Evidence for Carcinogencity*, edited by N.J. Wald and R. Doll, 1985, pp. 183–94.

35. The toxicokinetics of nitrosation are still not clearly understood. For example, Ohshima and Barsch found increases in nitrosoproline excreted when nitrate and proline (an amino acid) were ingested by a male volunteer. However, when the compounds were ingested separately, no increase in nitrosoproline was detected. Ohshima and Barsch, "Quantitative Estimation of Endogenous Nitrosation in Humans by Monitoring N-nitrosoproline Excreted in the Urine," *Cancer Research*, vol. 41, 1981, p. 3658.

36. *The Health Effects of Nitrate, Nitrite and N-nitroso Compounds*, National Academy Press, Washington, D.C., 1981. Also see R.L. Shank and P.N. Magees's synopsis in "Toxicity and Carcinogenicity of N-Nitroso Compounds," in *Myotoxins and N-nitroso Compounds*, vol. 1, CRC Press, Florida, 1981, p. 185.

37. F. J. Akin and A. E. Wasserman, "Effect on Guinea Pigs of Feeding Nitrosomorpholine and Its Precursors in Combination with Ascorbic Acid," *Food Chemistry and Toxicology*, vol. 13, 1975, p. 239.

38. Three British epidemiological studies have all shown no correlation or a negative correlation between nitrate exposure and cancer: David Forman, Samim Al-Dabbagh, and Richard Doll, "Nitrates, Nitrites, and Gastric Cancer in Great Britain," *Nature*, vol. 313, 1985, pp. 620–27; Shirley A. A. Beresford, "Is Nitrate in the Drinking Water Associated with the Risk of Cancer in the Urban U?" *International Journal of Epidemiology*, vol. 14, 1985, pp. 57–63; and al-Dabbagh et al., "Mortality of Nitrate Fertiliser Workers," *British Journal of Industrial Medicine*, vol. 43, 1985, pp. 507–15. An earlier study by M J. Hill et al. comparing stomach cancer mortality in two English towns found a relative risk of 1.9 for females in the population exposed to the higher levels of nitrate in drinking water. The study, however, did not control for socioeconomic disparities that

provide a more plausible explanation for the increased incidence. Also, nitrate concentrations of 90 mg/l found in the water of the exposed population are nine times higher than the existing and proposed standards. See M. J. Hill et al., "Bacteria Nitrosamine and Cancer of the Stomach", *British Journal of Cancer*, vol. 28, 1973, p. 562.

39. See U.S. EPA, *National Summary of State Water Quality Standards: Nitrates/Nitrites/Ammonia*, prepared by Nalesnik Associates Incorporated, Washington D.C., September 1980.

40. See SDWA, Sec. 1412(a)(1).

41. See *Federal Register*, vol. 50, November 13, 1985, p. 46936.

42. U.S. EPA, Office of Drinking Water, Criteria and Standards Division, *Final Draft for the Drinking Water Criteria Document on Nitrate/Nitrite*, prepared by ICAIR, Life Systems, Inc., Cleveland, Ohio, TR–540–59D, Contract 68–01–6750, October 1985 (hereinafter *Criteria Document*, 1985).

43. $\text{ADI} = \dfrac{(\text{NOAEL or LOAEL in mg/kg/day}) \ (\text{Body Wt in Kg})}{\text{Uncertainty/Safety Factor}} = \text{mg/day}$

AADI = ADI/Drinking Water Volume in l/day = mg/day

NOAEL = No observed adverse effect level

LOAEL = lowest observed adverse effect level

Drinking water volume = 2 l/day for adults, 1 l/day for children

It is important to note that EPA has recently changed this measure to the new RFD that measured mg/kg/per person/day.

44. *Drinking Water and Health*, vols. 1 and 6, National Academy of Sciences, Washington D.C., 1977 (Chapters 1–2) and 1986.

45. Graham Walton, "Survey of Literature Relating to Infant Methemoglobinemia Due to Nitrate Contaminated Water," *American Journal of Public Health*, vol. 41, 1951, pp. 986-95.

46. Ibid., pp. 989-90.

47. *Criteria Document*, pp. VIII 21–VIII22.

48. These are E. F. Winston, R. G. Tardiff, L. J. McCabe, "Nitrate in Drinking Water," *Journal of the American Water Works Association*, vol 63, 1971, pp. 96–98; V. W. Toussaint and K. Wurkert, "Metahämoglobinamie im Säuglingsalter," In *Nitrat-Nitrit-Nitrosamine in Gewassern*, edited by F. Selenka, Deutsche Forschungsgemeinschaft, Bonn, Germany, 1982, pp. 136–42; and W. E. Donahoe, "Cyanosis in Infants with Nitrates in Drinking Water as Cause," *Pediatrics*, vol. 3, 1949, pp. 308–11.

49. G. F. Craun et al., "Methemoglobin Levels in Young Children Consuming High Nitrate Well Water in the United States," *International Journal of Epidemiology*, vol 10, 1981, p. 309.

50. *Criteria Document*, 1985, p. VIII–22.

51. During the review by the SAB, it was pointed out by Herschel Griffin that the terms uncertainty factor and safety factor were used interchangeably within the document when they in fact are not synonymous. Insofar as data is not based on animal studies, the term "uncertainty factor" is actually a misnomer. The reduction to a level of 1 mg/l is actually a safety factor.

52. *Criteria Document*, 1985, p. VIII-23.

53. U.S. EPA, *National Academy of Sciences Drinking Water Criteria*, 73-033, Washington D.C., 1973.

54. U.S. EPA, Office of Planning and Standards, *Quality Criteria for Water*, Washington, D.C., 1976, p. 203.

55. *Federal Register*, vol. 50, no. 219, November 13, 1985, p. 46972.

56. The letter cited studies by the Cornell University Center for Environmental Research that calculated that if the average nitrate concentration in an area was 5.8 mg/l, then 10 percent of the samples from that area would exceed the 10-mg/l standard. The suggested revision to a 2-mg/l level would ensure a 99.9 confidence level of meeting the 10-mg/l standard. EPA's response did not address this argument at all but fell back on the previous toxicological information. The results from the Cornell study can be found in *The Long Island Comprehensive Waste Treatment Management Plan*, vol. 2, *Summary Documentation*, Chapter 5, "Nitrates," Project Director, Lee Koppelman, Nassau Suffolk Regional Planning Board, Hauppauge, New York, 1978, pp. 187–221.

57. Interview with Jennifer Orme, Office of Drinking Water, Health Effects Branch, July 25, 1988.

58. Interview with Dan Byrd; interview with Robert Tardiff, Environ Corporation, January 21, 1988.

59. *Transcript*, August 21, 1986, p. 13.

60. Ibid., pp. 11–12.

61. Report on EPA Science Advisory Board meeting, internal memo from Ruth Shearer to Life Systems, September 3, 1986.

62. Interview with Ken Bailey, January 21, 1988.

63. Memo from Shearer.

64. Carl J. Johnson et al., "Fatal Outcome of Methemoglobinemia in an Infant," *Journal of the American Medical Association*, vol. 257, 1987, pp. 2796–97. See also Carl J. Johnson, "Methemoglobinemia: Is It Coming Back to Haunt Us?" *Health and Environment Digest*, vol. 1, no. 12, January 1988, pp. 3, 8.

65. *Transcript*, August 21, 1986, p. 15.

66. See, for example, comments of Brubaker, *Transcript*, August 22, 1986, p. 139.

67. Interview with Richard Cothern, Executive Secretary, EHC, January 21, 1988.

68. Interview with Robert Tardiff.

69. *Transcript*, August 21, 1986, p. 38.

70. See comments of Marshall Johnson, ibid., p. 46.

71. *Transcript*, August 21, 1986, p. 60.

72. See comments of Dr. Nancy Kim, ibid., pp. 67–68. See also Laura C. Green et al., "Nitrates in Human and Canine Milk" (letter to the editor), *New England Journal of Medicine*, vol. 306, 1982, p. 1367.

73. See U.S. EPA, *Estimated National Occurrence and Exposure to Nitrate/Nitrite in Public Drinking Water Supplies,* revised draft, 1987.

74. *Transcript*, August 21, 1986, p. 9.

75. Cotruvo explained: "I think I would be happy if the committee were to take a document and concentrate on just the QTE section . . . which says, these are the half dozen experiments which we think are the most critical ones which we used to make the judgement. What's the significant toxicology? what's the significant endpoint? That's the part we want the scientists to tell us is right or wrong. Is it a reasonable interpretation or an unreasonable interpretation, missing a significant study or is it comprehensive?"

Transcript, August 21, 1986, 170–71. Ed Ohanian in an interview two years later still held this position.

76. *Transcript*, August 21, 1986, pp. 154–57.

77. Ibid. p. 113.

78. *Transcript*, August 22, 1986, p. 148.

79. Ibid., p. 150.

80. Ibid., p. 150.

81. Comments of Johnson, *Transcript*, August 22, 1986, p. 135.

82. For example, the absence of any recent research directly measuring methemoglobin levels was criticized for its implicit assumption that infant's diets have not changed since studies were made over thirty years ago.

83. See comments of Nancy Kim, *Transcript*, August 21, 1986, p. 79.

84. For example, ecological studies, case control studies, and epidemiological studies were mixed together. Creating a hierarchy of evidence beginning with the epidemiological studies was suggested.

85. Joe Cotruvo, Director of the Criteria and Standards Division, Office of Drinking Water: "They aren't necessarily intended to be comprehensive literature reviews. They're primarily intended to be assessment of the core literature that relates to the particular decision that we're making and the literature that makes a difference." *Transcript*, August 21, 1986, p. 168.

86. Ibid., p. 63.

87. "The only thing we have here is a disclaimer as to what it is not." Comments of Herschel Griffin, *Transcript*, August 21, 1986, p. 54.

88. See comments of Brubaker, *Transcript*, August 21, 1986, p. 117.

89. See comments by Dr. David Kaufman, p. 164, and Herschel Griffin, *Transcript*, August 21, 1986, p. 159–60.

90. *Transcript*, August 21, 1986, p. 5.

91. See comments of Marshall Johnson, *Transcript*, August 21, 1986, p. 178.

92. See comments of David Kaufman, *Transcript*, August 21, 1986, p. 168.

93. At a later point, Griffin cited the IARC as being similar to the WHO technical committee, both involving intense sessions of "hands-on preparation" by the review committee. *Transcript*, August 22, 1986, p. 7.

94. Ibid., p. 39.

95. Ibid., p. 119.

96. *Transcript*, August 22, 1986, p. 21.

97. Ibid., p. 22.

98. Telephone interview with Gary Carlson, Chairman of the Safe Drinking Water Subcommittee, June 1, 1988.

99. See *Transcript*, August 22, 1986.

100. Interview with Robert Tardiff.

101. In doing so the subcommittee implicitly sided with Dr. Shearer's objections over the agency's position.

102. Interview with Ken Bailey.

103. U.S. EPA, Criteria and Standards Division, *Final Draft for the Drinking Water Criteria Document on Nitrate/Nitrite*, May 1987, pp. VIII 20-VIII 21. Surprisingly, another justification for use of the Walton study included the claim, "The study is an epidemiology study and thus presumably included infants in all stages of health and sickness." Ibid., p. VIII 21.

104. Interviews with Ken Bailey and Ed Ohanian, January 21, 1988.

105. Interview with Ed Ohanian.

106. Interview with Ken Bailey.

107. Interview with Ed Ohanian.

108. Interview with Joseph Cotruvo, Director, Division of Criteria and Standards, Office of Drinking Water, March 18, 1988.

109. Telephone interview with Ken Bailey, September 7, 1988.

110. Interviews with Ken Bailey and Ed Ohanian, January 21, 1988.

111. Interview with Dan Byrd.

112. Interview with Robert Tardiff.

113. Interview with Dan Byrd.

114. Anna M. Fan, Calvin C. Willhite, and Steven A. Book, "Evaluation of the Nitrate Drinking Water Standard with Reference to Infant Methemoglobinemia and Potential Reproductive Toxicity," *Journal of Regulatory Toxicology and Pharmacology*, vol. 7, 1987, pp. 135–48.

115. Proposed standard notice for *Federal Register*, January 1989, p. 40 CFR part 141, 142, 143.

116. Ibid., p. 76.

117. Ibid, p. 77.

118. Ibid., p. 78.

119. Margaret Dorsch et al., "Congenital Malformations and Maternal Drinking Water Supply," *American Journal of Epidemiology*, vol. 119, 1984, pp. 473–86; telephone interview with John Davidson, Policy Analyst, OPPE, EPA, September 7, 1988.

120. Telephone interview with Ken Bailey.

121. Interview with John Sheets, Director of Analytical Laboratories, North Carolina Department of Public Health, February 11, 1988.

122. Interview with Michelle M. Frey, Environmental Engineer, Assistant Director of the Office of Government Affairs, American Water Works Association, March 17, 1988.

123. See Mohamed F. Dahab and Young Woon Lee, "Nitrate Removal from Water Supplies Using Biological Denitrification," *Journal of the Water Pollution Control Federation*, vol. 60, 1988, p. 1670.

124. Interview with Joe Hollstein, Director of Public Works, Ceres, California, May 31, 1988.

125. Interview with Jeanne W. Appling, Toxicologist, Center for Environmental Research, Cornell University, April 29, 1988.

126. Interview with Keith Porter, Director of the Water Resources Institute, Cornell University, April 29, 1988.

127. Interview with Robert Tardiff.

128. Interview with Michelle M. Frey.

129. Interview with Dan Byrd.

130. Interview with Joseph Cotruvo.

131. Interview with Michelle M. Frey.

132. Interview with Gary Carlson.

133. Interview with Robert Tardiff.

134. Interview with Jay Donald Johnson, UNC School of Public Health, February 15, 1988.

135. Interview in 1988.

136. Richard M. Cooper, Attorney, Williams and Connolly, comments at the Belmont Workshop, October 26, 1988.

137. Interview with David Kaufman, UNC Department of Pathology, February 12, 1988.

138. Interview with Dan Byrd.

139. Interview with Nancy Kim, April 28, 1988.

140. Interview with David Kaufman.

141. Interview with Robert Tardiff.

142. Interview with Gary Carlson.

143. Comments by Terry R. Yosie.

144. Interview with Gary Carlson.

145. Interview with Robert Tardiff.

146. Interviews with Daniel Menzel. Also comments of Robert Tardiff, Gary Carlson.

147. Interview with Robert Tardiff.

148. Ibid.

149. Interview with Joe Cotruvo.

150. Interview with Ed Ohanian.

151. Interview with Gary Carlson.

152. Interview with Ed Ohanian.

153. Interview with Ken Bailey.

154. Judicial deference to external scientific review is discussed in several contexts in Jasanoff, *The Fifth Branch of Government* (Cambridge, 1990), Chapter 4.

155. Interview with Joseph Cotruvo.

156. Interview with Ken Bailey.

157. Interview with Robert Tardiff, January 21, 1988.

158. Interview with Dan Byrd.

159. Interview with Nancy Kim.

160. Ibid.

161. Interview with Joseph Cotruvo.

162. Interview with David Kaufman.

CHAPTER 9

CARBON MONOXIDE

John D. Graham and David Holtgrave

The Health Effects Institute was just a few years old when its sponsors, EPA and the auto industry, requested that it investigate the effect of acute exposure to carbon monoxide on the health of cardiovascular patients. It is hard to imagine a problem that could be better tailored for HEI: a well-defined scientific question that was of profound significance to the setting of EPA's national air quality goal for carbon monoxide. Previous studies of the question had been plagued by weak study design and suspicions of data "fudging."

EPA had been struggling in vain for fifteen years to find a scientific basis for its CO standard in the face of persistent criticism from Detroit. As the major man-made source of emissions in the country, automobile manufacturers had grown weary of being the victim of shoddy science. They were prepared to accept a tough air quality goal for CO if, and only if, it was justified by impeccable science.

In this chapter we trace the checkered history of EPA's CO standard and describe HEI's effort to resolve the controversy. While it is too early to make a definitive evaluation of HEI's contribution, our interviews uncovered broad, if not universal, support for the proposition that HEI delivered a landmark study in regulatory science. The chapter explores why the HEI study of CO has received such high marks and how it might serve as a model of regulatory science.

EPA'S SHAKY SCIENTIFIC FOOTING

One of EPA's key responsibilities under the 1970 Clean Air Act Amendments was to establish a national ambient air quality standard (NAAQS) for carbon monoxide. On April 30, 1971, EPA promulgated a health-based ("primary") CO standard at levels of 9 ppm, eight-hour average, and 35 ppm, one-hour average. EPA argued that this standard would satisfy the legislative directive to "protect the public health with an adequate margin of safety."[1]

The scientific basis of the standard came from studies suggesting that elevated carboxyhemoglobin (COHb) levels in the range of 2 to 3 percent could damage the central nervous system. Although questions were raised about the validity of these studies, EPA was particularly impressed by evidence that CO exposure could impair a person's ability to discriminate intervals of time.[2] By way of comparison, adults exposed to 9 ppm CO for eight hours reach 0.6 percent to 1.0 percent COHb at rest and about 1.4 percent COHb after moderate activity. One-hour exposures to 35 ppm CO generate COHb levels of about 2 percent after moderate activity. Since the neurobehavioral studies had detected effects at COHb levels as low as 1.8 percent, some scientists were concerned that the promulgated standard did not contain an "adequate margin of safety."

During the 1970s the neurobehavioral studies came under intense scrutiny by EPA, industry, and academic scientists. When EPA began a review of the CO standard in 1978, it concluded that the alleged psychomotor effects could no longer be used to justify the standard. While there had always been concern about the reliability of the subjective responses of subjects, the decisive problem was the inability of later studies to replicate such effects at COHb levels below 5 percent.[3] As a result, EPA searched for an alternative scientific basis of its 1971 standard.

THE ARONOW FLAP

On August 18, 1980, EPA proposed to retain the eight-hour standard of 9 ppm CO and tighten the one-hour standard of 35 ppm. The proposal was based primarily on a new line of scientific research: several clinical experiments purported to demonstrate that exercising heart patients experienced a faster onset of angina when their COHb levels were elevated into the range of 2 to 3 percent. All but one of the clinical studies cited by EPA were conducted under the leadership of Wilbert S. Aronow, a scientist at the Veterans Administration who had been funded in part by EPA's Office of Research and Development.[4]

While the quality of some of Aronow's early work on CO was questioned in letters to the editor of the *Annals of Internal Medicine*, many respected scientists were convinced that the effect Aronow was reporting was biologically plausible. EPA's 1980 proposal was made with the approval of the agency's Clean Air Scientific Advisory Committee (CASAC), a panel of independent scientists.

After taking public comment on the proposal, EPA saw its CO review delayed several years by the transition from President Carter to Reagan and the tenure of Anne Gorsuch at EPA. Nonetheless, by early 1983 Gorsuch was nearing a decision that would have set the NAAQS for CO at a level to prevent COHb levels from exceeding 3 percent.[5]

To the surprise of EPA officials, The *Washington Post* reported in early 1983 that some of Aronow's research on experimental drugs had been the subject of investigations by the Veterans Administration and the Food and Drug Administration. Concerns had been raised about falsification of results reported from Aronow's laboratory. EPA learned that the VA had directed Aronow in 1980 to discontinue all research activities, and in 1982 FDA had limited Aronow's right to serve as a clinical investigator for pharmaceutical companies.[6]

Although EPA had no reason to doubt the authenticity of Aronow's work on CO, the agency was about to promulgate a national air quality goal that would be based primarily on his data. The agency decided to establish an independent peer review committee to evaluate Aronow's studies and audit his documentation. On the basis of a careful review that included visits to Aronow's laboratory, the committee, which was chaired by Steven Horvath, a nationally recognized CO expert, recommended that EPA not rely on Aronow's studies when setting the standard.[7] Although Aronow vigorously contested the findings of the Horvath committee, EPA decided that it would have to ignore his data, except perhaps as a factor to keep in mind when setting a margin of safety in the standard. EPA staff issued an addendum to the criteria document on CO and prepared a new staff paper for the EPA administrator in light of the loss of the Aronow data.[8]

In the midst of this and other controversies, William Ruckelshaus became administrator of EPA, and Morton Lippmann of New York University was appointed chair of the CO subcommittee of the agency's Clean Air Scientific Advisory Committee. When all the dust had settled, EPA staff and Lippmann's subcommittee were left with only one reference that could support the existing standard. A 1972 study by E. W. Anderson and coworkers had examined ten exercising heart patients and found a significant reduction in (self-reported) time to onset of angina at COHb levels in the range of 2.9 to 4.5 percent.[9] It was this study that had advanced the hypothesis that Aronow attempted to prove. The Anderson study was small, unreplicated, and outdated and found the same amount of effect at both COHb levels.

In consultation with Lippmann's subcommittee, EPA decided in 1985 to retain the current standard in spite of its limited scientific basis. EPA argued that the large degree of uncertainty created by the lack of data should lead to a large margin of safety in the final standard. CASAC ultimately approved this line of reasoning.[10] In the published decision supporting the retained standard, EPA emphasized that several new studies were under way, including one by HEI, that should provide more and better data for the next review of the standard. When Lee Thomas succeeded Ruckelshaus as EPA administrator in 1985, he was told that the CO standard would be revised as soon as the several studies in progress were completed.[11]

EPA's decision was baffling to industrial scientists and to Congressman John Dingell, the industry's ally and chair of the House Committee on Energy and Commerce. At congressional hearings Lippmann attempted to explain to Dingell why it was that the removal of the Aronow data could still permit retention of the current standard. To put it mildly, Dingell was not impressed with the "margin of safety" rationale that EPA and Lippmann had embraced.[12]

ENTER THE HEALTH EFFECTS INSTITUTE

As noted in Chapter 4, the Health Effects Institute has a unique organizational structure that places much of its authority in the hands of two standing committees of scientists. The Research Committee is responsible for charting research directions, fashioning requests for proposals, deciding which proposals are worthy of financial support, and overseeing the progress of the studies. The Review Committee evaluates the work that HEI has sponsored and publishes a critique of each completed study. Hence the Review Committee plays no role in planning or funding the research it evaluates. Both committees are comprised primarily of academic scientists who serve as consultants to HEI's small in-house staff.

From the beginning, HEI's Research Committee established a reputation for making independent decisions about what research topics would be given priority and which investigators would be funded. HEI's sponsors were often frustrated by the Research Committee's independence, particularly since the committee often took a more long-term view of HEI's research mission than the sponsors might have preferred. When the sponsors requested that HEI investigate the effects of CO exposure on heart patients, it was not a foregone conclusion that the HEI Research Committee would grant this request.

The HEI Research Committee utilizes an investigator-initiated style of research rather than a style that resembles contract work for HEI. Moreover, some HEI scientists are convinced that basic research into the mechanisms of pollution-induced health effects offers more promise than strictly empirical investigations of dose-response relationships. Understanding mechanisms is important to understanding the significance of a health effect and in developing better methods for assessing dose-response relationships. At the same time, HEI's Research Committee knows that it has to be somewhat responsive to the perceived needs of its sponsors, especially on issues of regulatory importance. Hence HEI's program has become a mix of basic and applied regulatory studies.

In the case of carbon monoxide, the HEI Research Committee seized on the sponsor requests as an opportunity to demonstrate that HEI could deliver first-rate science on a question of critical near-term regulatory significance.[13] Under the leadership of Chairman Walter Rosenblith, HEI's Research Committee decided to examine the CO question in a more hands-on fashion than is customary.

Rather than begin by issuing a request for proposals, the Research Committee created an Advisory Committee that helped plan and monitor the study. Chaired by John W. Tukey of Princeton University, the Advisory Committee designed a preliminary study plan with the aim of answering the same fundamental question that Aronow had explored. However, the HEI plan contained provisions for participation by multiple centers, an expanded sample size, and both objective and subjective measures of time to onset of angina.

The Advisory Committee began the process of recruiting centers to participate in the study by submitting the preliminary protocol to several hundred researchers with relevant expertise. The search process led to the selection of investigator teams at three test centers: the Francis Scott Key Medical Center (Johns Hopkins University), the Rancho Los Amigos Medical Center, and the St. Louis University School of Medicine. A detailed protocol for the study was devised by the investigators, working in conjunction with HEI's staff and advisory committee.

The logistical plans for the study were not simple. Each test center was to submit blood samples to the Reference Laboratory at the St. Louis University Medical Center, where COHb levels would be determined by the most accurate method, gas chromatography (GC). All data collected at the test centers would be submitted to the Harvard School of Public Health's Statistical and Data Management Center for analysis. At Tukey's insistence, the study team embraced a detailed plan of analysis before the data were collected. A Quality Assurance Team from Arthur D. Little, Inc., was responsible for visiting the test centers and making sure that each investigator conducted the study according to protocol. Jane Warren of the HEI staff served as project manager for the study and played a critical coordinating role.

UNEXPECTED SNAGS IN THE HEI STUDY

Although EPA requested that HEI launch such a study in July 1983, the study did not actually begin until May 1985. While part of this delay was attributable to the planning activities, the study team also experienced delays in ironing out the protocol. One of the biggest challenges was to agree on the exposure regimen, which required a pilot study to produce data that would inform the final exposure regimen.

The HEI investigators knew that any fixed CO exposure regimen would produce varying COHb levels due to differences among subjects in rates of CO uptake. Since the objective of the study was to assess cardiac effects in groups of subjects at 2 percent and 4 percent COHb, it was critical that each subject's COHb level be controlled precisely.

After the pilot study, the team decided to focus on obtaining identical COHb targets for each patient. Targets would be achieved by adjusting the CO exposure regimen for each patient based on experimentally determined uptake rates. Working out the details of this rigorous exposure protocol led to unexpected delays in the study as well as unanticipated costs. At this stage of the investigation, no one really knew how long the study would take or how much it would cost.

Once the final protocol was written, each center began the task of recruiting subjects into the study. Recruitment of subjects proved to be much more difficult and time consuming than the investigators had anticipated. The selection criteria called for male patients who did not smoke and who had objective evidence of coronary artery disease and angina, including evidence of reproducible, exercise-induced ischemic (ST) changes on the electrocardiogram (ECG) during exercise treadmill testing. The protocol included eight specific inclusion criteria and specific exclusion criteria that were used by each of the test centers. In the final analysis a sample of sixty-three patients—divided roughly equally between the three centers—completed the study.

Each subject was evaluated on four separate visits, once for qualification purposes plus three double-blinded visits in accordance with the exposure protocol. Once the data were collected, analyzed, and written up in a draft final report, the HEI Review Committee performed a rigorous peer review of the entire effort. The Review Committee's evaluation of the study was published with the HEI final report. While many scientists regard HEI's review process as a great virtue, it does add time and cost to the process. Some scientists believe that it is inappropriate for HEI to publish a critique of every supported study, an activity that might be perceived as second-guessing the work of funded investigators.[14]

The normal procedure at HEI is to publish a final report as soon as the Review Committee's work is completed. In this case significant delays in release of the HEI report occurred while the authors wrote a condensed version of the HEI report for submission to the *New England Journal of Medicine*, a prestigious, peer-reviewed scientific journal. Initially, the HEI scientists were uncertain whether the editors of the journal would be interested in publishing a paper that had only marginal relevance to the practice of medicine. Later, the peer review process at the journal led to prolonged delays in publication.

The publication delays were particularly irritating to EPA officials, who in early 1989 were beginning their much-anticipated review of the NAAQS for carbon monoxide. Even though EPA officials obtained from HEI a confidential draft copy of the report, the agency's work on a criteria document is supposed to be open to the public. It is very difficult to write a criteria document when everyone involved does not have access to the same information.

Looking back on the entire multicenter study, one participant in the HEI process characterized the effort as "something like a military procurement process; no one really knew how long the study would take or how much it would ultimately cost."[15] In the final analysis the study cost $2.5 million, and the time period from when the study began to when the report was publicly available was five years. While the degree of care and rigor in the HEI study was unprecedented in environmental science, a stiff price was paid in both dollars and time.

HEI'S SCIENTIFIC COMPETITORS

When the Aronow flap undercut the scientific rationale for EPA's CO standard, HEI was not the only scientific group to move into the vacuum. At an EPA laboratory affiliated with the University of North Carolina, David S. Sheps and colleagues had moved quickly to explore the validity of Aronow's hypothesis. Sheps studied thirty nonsmoking patients with heart disease defined by exercise-induced ST depression. In a randomized, double-blind protocol, the patients were exposed to clean air (COHb = 1.5 percent) and to 100 ppm CO (COHb = 3.8 percent). The most innovative feature of this study was the use of left ventricular ejection fraction (EF) obtained from rest-exercise radionuclide studies in addition to the more traditional information about time to onset of angina reported by patients.

The interpretation of the results of the Sheps experiment at 3.8 percent COHb proved to be quite controversial. His scientific paper was entitled "Lack of Effect of Low Levels of Carboxyhemoglobin on Cardiovascular Function in Patients with Ischemic Heart Disease."[16] The authors emphasized that exposure to CO produced no significant change in time to onset of angina. Although the EF measure was affected by CO exposure, Sheps and his colleagues dismissed this effect as "minimal" and described their results, unlike Aronow's, to be negative.

EPA's assistant administrator of research and development, Bernard Goldstein, was not convinced that Sheps and his colleagues had properly interpreted their data. Goldstein was concerned that the Sheps interpretation was flawed by a classic type II error: saying that no effect exists when in fact an effect does exist. Goldstein pointed out that the most innovative feature of the study, the change in EF, did appear to be influenced by COHb levels around 3.8 percent. Despite repeated conversations with Sheps, Goldstein and his staff were not successful in persuading him to alter his interpretation of the data.[17]

Sheps decided instead to repeat the same type of experiment with COHb levels around 6 percent. In this experiment he found some evidence of a CO effect on angina. From a scientific perspective, this was the first demonstration of an effect of CO on change in EF.[18] However, the results did not provide EPA any scientific basis for its shaky CO standard because the COHb levels among tested subjects were too high.

The California Air Resources Board (CARB) also moved into the vacuum by commissioning scientists at the University of California at Irvine to investigate the Aronow hypothesis. The study, directed by Michael T. Kleinman and James L. Whittenberger, was launched after the HEI study began but was completed before HEI's more cumbersome multicenter team had finished data collection. Kleinman and Whittenberger chose a target COHb level of 3 percent so that the results would be directly relevant to the existing NAAQS for CO.

Unlike the first Sheps experiment, the CARB study found rather clear evidence that COHb levels of 3 percent are responsible for reduced exercise tolerance in people with angina. It reported CO-induced reductions in self-reported time to onset of angina (-6.9 percent), although some of the measures of exercise tolerance were not significantly affected by CO. The effects of CO were found to be especially large among those nineteen patients of mass 95 kg and below. In summary, Kleinman and Whittenberger reported to CARB in late 1985 that their findings "are in agreement" with those of Aronow.[19]

In light of the apparently conflicting results of the Sheps and CARB studies, interest in the results of the ongoing HEI investigation was heightened. The unexpected delays in the HEI study further contributed to both the suspense and the frustration among everyone who was anticipating EPA's reconsideration of the NAAQS for CO.

HEI'S FINDINGS

As noted earlier, the HEI multicenter study included sixty-three nonsmoking male subjects with stable angina. The effects of both 2 percent and 4 percent COHb were tested. Two methods of measuring COHb, gas chromatography and CO-oximeter, were used. The primary endpoints examined were time to onset of ischemic ST-segment change (via electrocardiogram) and time to onset of angina (self-reported).

The investigators reported 5.1 percent and 12.1 percent decreases in the time to onset of ST-segment changes at 2 percent and 4 percent COHb, respectively. They also reported 4.2 percent and 7.1 percent decreases in the self-reported time to onset of angina. These percentage changes were calculated relative to a control group of patients exposed to clean air.[20] The percentage effects were statistically significant using a one-tailed test of significance. The HEI investigators and Review Committee interpreted these results as generally supportive of the hypothesis originally advanced by Anderson and tested by Aronow.

Public release of the HEI findings was delayed for over a year while the report was under consideration at the *New England Journal of Medicine*. This journal does not accept papers that have already been released to the public. The delay was quite irritating to people interested in EPA's anticipated review of the CO standard, since the review process would need to consider the HEI methods and findings. While HEI did release several confidential copies of the report to EPA for working purposes prior to publication, the report was not actually published in the *New England Journal of Medicine* until December 1989.[21] Accompanying the published study was a rather provocative editorial suggesting that the adverse health effects associated with CO might not exhibit any no-effect level.[22]

ANTICIPATED IMPACT ON EPA'S CO STANDARD

Taken together, the UNC, CARB, and HEI studies provided significant new evidence that is relevant to setting EPA's air quality standard for CO. We interviewed eleven science-policy experts familiar with EPA's standard-setting process in order to determine what the impact of this new evidence is likely to be. After briefing each expert on the three studies and EPA's upcoming review process, we asked the experts to assess which study would have the most influence on the final standard. We also asked each expert to assess the likelihood that the standard would be changed (tightened or relaxed) in light of the new evidence.[23]

The consensus of the experts was that all three studies would be considered by EPA and taken into account when the CO standard was reviewed. While each study was considered a significant contribution to the overall weight of the evidence, the experts tended to believe that the HEI study would receive the greatest weight. The explanation for the HEI study's greater predicted weight lay not primarily with characteristics of the institute per se but rather in several specific technical features of the study. In particular, the HEI study was distinguished by (1) the use of multiple centers, (2) a relatively large sample size of patients, (3) the objective measure of time to onset of myocardial ischemia, (4) a rigorous quality-assurance program, (5) a more accurate procedure (GC) for measuring COHb levels, and (6) a data analysis plan that was laid out in advance prior to data collection. Few studies in the history of environmental science have been carried out with this kind of rigor and care. Taken together, these features of the study may cause it to be considered a landmark contribution to regulatory science.

Taking into account all the new evidence, we asked each of the eleven experts to assess how likely it is that the standard will be tightened, retained, or relaxed. The same question was then posed again under the hypothetical assumption that the HEI results would not be available. In this way we intended to assess whether the HEI findings per se are likely to have a significant regulatory impact.

The judgmental probability assessments are presented in Table 9.1. The expert participants were assured anonymity to encourage candid responses to questions. Although there was some variability in expert opinion, the predominant opinion was that the standard is likely to be retained. If the standard is changed, it is more likely to be tightened than relaxed. If the standard is tightened, the one-hour standard is more likely to be tightened than the eight-hour standard. A minority of the experts, however, expressed the opinion that it is more likely than not that the standard will be tightened. Several experts commented that the probability of a tighter standard will depend not just on the science per se but on the degree of advocacy work that environmental organizations choose to devote to CO.

Interestingly, the experts believed that the CARB and UNC studies together (without the HEI findings) would lead to roughly the same standard-setting outcome. At most, the HEI findings have made it slightly more likely that the standard will be tightened than would have been the case without the HEI findings. Recall that the CARB findings showed significant effects on angina patients at 3 percent COHb, which is close to the levels (2 percent and 4 percent COHb) used by the HEI investigators. Overall, the major effect of the CARB and HEI findings is primarily to strengthen the scientific basis of the existing standard.

CONCLUSION

HEI's multicenter study of carbon monoxide was the institute's first opportunity to produce original data that are likely to serve as the centerpiece of a standard-setting deliberation at EPA. Since EPA's review of the CO standard is not completed, it is premature to draw any conclusions about how decisive HEI's contribution may prove to be. The early signs suggest that although the HEI study may not cause a dramatic change in the standard, it will provide the primary scientific basis for the existing standard that has heretofore been sorely lacking.

Table 9.1
Judgmental Assessments of the Probability of a New CO Standard With and Without Consideration of the HEI Study

Effect on CO Standard	A	B	C	D*	E	F	G	H*	I	J	K	Mean
With HEI												
Tighten	10 (20)	20	29 (65)	58	50	0	30	17	10	50 (65)	30	28
Retain	85 (75)	80	70 (34)	37	50	95	60	79	85	50 (35)	70	69
Relax	5 (5)	0	1 (1)	5	0	5	10	4	5	0 (0)	0	3
Without HEI												
Tighten	10 (10)	20	5 (39)	55	20	0	20	10	5	50 (65)	20	20
Retain	85 (85)	80	65 (60)	35	50	80	70	80	85	50 (65)	80	69
Relax	5 (5)	0	30 (1)	5	30	20	10	10	10	0 (0)	0	11

Notes: The symbol "*" indicates that the numbers reported are midpoints of ranges provided by the experts. Experts A, C, and J provided different probability assessments for the eight-hour and one-hour standards. Their assessments for the one-hour standard are presented in parentheses. "Mean" is a group mean.

In reviewing HEI's work on CO, it is apparent that the process was slower, more costly, and more cumbersome than some people anticipated at the beginning of the study. The work at both CARB and UNC—conducted under more traditional institutional arrangements—proceeded more quickly and led to significant and relevant results. When all is said and done, however, the HEI multicenter study may fairly be characterized as a landmark investigation because of the extraordinary steps taken to ensure the quality of the data, the appropriateness of the data analysis, and the proper interpretation of the study results.

One might ask whether the multicenter model of regulatory science could be applied fruitfully to other air pollutants or other environmental problems. In principle, other applications are certainly possible. It should be emphasized, however, that the CO issue was well tailored for HEI because the scientific hypothesis had already been framed and the standard-setting process was begging for a test of this hypothesis. The research questions around pollutants such as ozone and particulates are perhaps less structured, which might make it difficult to expect such a powerful result from a single multicenter study.[24] Even in the case of CO there were some differences between the results at the participating centers.

The more significant lesson of the CO experience is that the HEI model can work to produce first-rate environmental science. While HEI does not have significant in-house capability to produce original data, the CO case reveals that HEI has the ability to recruit qualified scientists to participate in a complicated, large-scale investigation. Moreover, this was a study that HEI produced in direct response to requests from its sponsors. Now that the study is completed, it is difficult to imagine that the data will ever suffer from the credibility problems that plagued the Aronow studies or the original research on psychomotor effects of CO exposure.

The CO experience does raise some questions about the costs and timeliness of the HEI process. Since the distinctive feature of the HEI study of CO was the rigorous protocol employed at multiple centers, it must be expected that the plans for such a study will be somewhat expensive and time consuming. Scientific quality comes at a significant price in dollars and time. One can only speculate about whether the study itself could have been conducted more swiftly and at lower cost.

On the other hand, the HEI sponsors and the public have a right to question whether publication delays should be tolerated to permit publication in a particular journal. The HEI report was already subjected to extensive peer review by the HEI Review Committee before it was submitted to the *New England Journal of Medicine*. Many first-rate scientific journals do not have such restrictions on public release. Obviously, the participating scientists may seek the academic prestige afforded by publication in this particular journal. (Defenders of HEI argue that the *New England Journal of Medicine* was the best publication due to its wide circulation.) In the case of regulatory science, however, certain academic niceties should arguably be compromised to serve the public interest.[25]

In the future, the leadership of HEI may wish to consider whether a different publication strategy would better address the need for timely publication.

NOTES

The authors gratefully acknowledge personal interviews with Richard Bates, Steven Colome, Robert Frank, Bernard Goldstein, Ian Higgins, Morton Lippmann, Roger McClellan, James McGrath, Frank Speizer, Lee Thomas, and Terry Yosie. Useful background materials were also provided by Thomas Grumbly, Bruce Jordan, Richard Paul, Harvey Richmond, Ken Sexton, Andrew Sivak, Lawrence Slimak, and Jane Warren.

1. *Federal Register*, vol. 36, 1971, p. 8186.

2. R. P. Beard and G. A. Wertheim, "Behavioral Impairment Associated with Small Doses of Carbon Monoxide," *American Journal of Public Health*, vol. 57, 1967, pp. 2012–22.

3. *Federal Register*, vol. 45, 1980, p. 55066.

4. W. S. Aronow and M. W. Isbell, "Carbon Monoxide Effect on Exercise-induced Angina Pectoris," *Annals of Internal Medicine*, vol. 79, 1973, pp. 392–95; W. S. Aronow et al., "Effect of Freeway Travel on Angina Pectoris," *Annals of Internal Medicine*, vol. 77, 1972, pp. 669–76; W. S. Aronow et al., "Aggravation of Angina Pectoris by Two Percent Carboxyhemoglobin," *American Heart Journal*, vol. 101, 1981, pp. 154–57.

5. Terry Yosie, former Director, EPA Science Advisory Board, personal communication, 1989.

6. Sheila Jasanoff, *The Fifth Branch of Government*, Harvard University Press, Cambridge, Mass., 1990, Chapter 11.

7. S. M. Horvath et al., letter to Lester D. Grant, U.S. Environmental Protection Agency, June 4, 1983.

8. U.S. Environmental Protection Agency, *Review of the NAAQS for Carbon Monoxide: Reassessment of Scientific and Technical Information*, EPA–450/5–84–004, July 1984, p. 13.

9. E. W. Anderson et al., "Effect of Low-Level Carbon Monoxide Exposure on Onset and Duration of Angina Pectoris: A Study of 10 Patients with Ischemic Heart Disease," *Annals of Internal Medicine*, vol. 79, 1973, pp. 46–50.

10. *Federal Register*, vol. 50, 1985, p. 37484.

11. Lee Thomas, former Administrator of EPA, personal communication, 1989.

12. Morton Lippmann, former chair, CASAC CO Subcommittee, personal communication, 1989.

13. Charles Powers, former Executive Director, Health Effects Institute, personal communication, 1989.

14. Frank Speizer, Professor, Harvard School of Public Health, personal communication, 1989.

15. Roger McClellan, member, HEI Research Committee, personal communication, 1989.

16. D. S. Sheps et al., "Lack of Effect of Low Levels of Carboxyhemoglobin on Cardiovascular Function in Patients with Ischemic Heart Disease," *Archives of Environmental Health*, vol. 42, 1987, pp. 109–16.

17. Bernard Goldstein, former Assistant Administrator for Research and Development, Environmental Protection Agency, personal communication, 1989.

18. K. F. Adams et al., "Acute Elevation of Blood Carboxyhemoglobin to 6% Impairs Exercise Performance and Aggravates Symptoms in Patients with Ischemic Heart Disease," *Journal of the American College of Cardiology*, vol. 12, 1988, pp. 900–909.

19. M. T. Kleinman and J. L. Whittenberger, "Effects of Short-Term Exposure to Carbon Monoxide in Subjects with Coronary Artery Disease," Final Report, California Air Resources Board, CARB Contract Number A3–081–33, November 26, 1985.

20. Health Effects Institute, *Acute Effects of Carbon Monoxide Exposure on Individuals with Coronary Artery Disease*, Cambridge, Mass. 1989.

21. E. N. Allred et al., "Shortterm Effects of Carbon Monoxide Exposure on the Exercise Performance of Subjects with Coronary Artery Diseases," *New England Journal of Medicine*, vol. 321, 1989, pp. 1426–1432.

22. Ibid.

23. J. D. Graham and D. Holtgrave, "Predicting EPA's Forthcoming CO Standard in Light of New Clinical Evidence," Final Report, Motor Vehicle Manufacturers Association, Detroit, Michigan, 1989.

24. Morton Lippmann, former chair, CASAC CO Subcommittee, personal communication, 1989.

25. Michael Walsh, consultant, personal communication, 1989.

CHAPTER 10

RESOLVING THE REGULATORY SCIENCE DILEMMA

John D. Graham

Regulators of toxic chemicals need knowledge to make competent decisions and the prestige of science to defend their decisions from attack. These needs militate in favor of the development of "regulatory science," that is, a close institutional relationship between science and regulatory power. Yet the norms of environmental regulation are at odds with some of the basic norms of science.

Regulators often must act swiftly in the face of uncertainty while permitting participation by nonexpert citizens who may have strong interests in the outcome. In contrast, first-rate scientists are reluctant to draw conclusions in the face of uncertainty. They place themselves in environments that foster objectivity and minimize political conflict, and they emphasize specialized expertise as a precondition to solving complex problems. Moreover, the public has delegated regulatory power to government officials who are ultimately accountable to the public for their actions. Scientists are accountable to the standards of their disciplines, not to the public at large. These considerations discourage a close institutional relationship between science and regulatory power.

In light of these considerations, how should society resolve the "regulatory science" dilemma in the context of controlling human exposure to potentially toxic substances? Is it possible to create a regulatory process that is both scientifically competent and politically accountable? This book has examined three institutional models that are relevant to answering these questions: the EPA Science Advisory Board, the Chemical Industry Institute of Toxicology, and the Health Effects Institute. This final chapter assesses how well these relatively new organizations have functioned and how they might function better in the future. It also speculates about how current trends in environmental policy are likely to affect the need for such institutional innovations in the future.

THE EPA SCIENCE ADVISORY BOARD

Of the three organizations considered in this book, the EPA Science Advisory Board is certainly the most secure. The board has grown rapidly in the last ten years, and no one has seriously suggested that it is not necessary. Many EPA managers have learned to see SAB ''approval'' as a necessary hurdle and, at times, as an asset in efforts to shepherd favored rules or projects through the EPA bureaucracy.[1] If anything, the board faces the challenge of setting its own priorities in the face of increasing demands from offices within EPA.

Historically, SAB has been primarily a reactive force, responding to issues raised by the administrator and program officials. Some observers believe that SAB is too permissive in letting the agency determine what issues will be addressed and what the scope of its reviews will be.[2] SAB has not ignored this criticism. For example, the agency's Office of Research and Development has not always been eager to receive SAB's advice about how to spend its money. In recent years SAB has been taking a hard look at the agency's research programs and suggesting new directions and institutional reforms. This new line of work should be encouraged because SAB's Executive Committee is in a unique position to see inconsistencies between EPA's operating needs and its research priorities.

In the future SAB should become a more proactive force in setting the nation's agenda for regulation and research. A strong case can be made that EPA is not addressing the environmental problems that pose the most serious threats to human and ecological health. Indeed, the agency's recent *Unfinished Business* report found that EPA priorities appear to be more responsive to public opinion than to scientific evidence and judgments.[3] Although it would not be desirable for EPA to ignore public opinion, SAB could play a stronger role in prodding Congress and the agency's leadership to focus on the more serious environmental problems. In recent years SAB has begun to make this type of contribution by helping place lead, radon, and global warming on the national agenda. Moreover, SAB has also made some far-reaching recommendations to expand and reorient the federal government's long-term research program on environmental questions.[4] To fulfill this role, SAB must not be too bashful about crossing the boundary between science and policy.

The Executive Committee of the SAB should be encouraged to address more of these ''big-picture'' questions, even though such projects are risky and at times controversial. The potential benefits of such projects are enormous, and SAB is strong enough to survive some misguided endeavors in this direction.

SAB's strength stems mainly from success at its bread-and-butter task of reviewing agency initiatives. The case studies reveal that SAB review does not have any predictable influence on regulatory outcomes. In the case study of nitrates, SAB and an EPA contractor pushed the agency in the direction of being more protective of public health than the agency might otherwise have been. In the case study of perchloroethylene, SAB acted as a restraining force when the agency seemed to be overinterpreting the available carcinogenicity data. Sometimes, as in the case of gasoline vapor, agency officials will attempt to use SAB approval to strengthen their case for regulatory action. Often, SAB will provide a stamp of approval to EPA's scientific analysis after one or more rounds of peer review and revision. The case study of formaldehyde provides a good illustration of a careful review that resulted in ultimate approval of the agency's risk assessment. The fact that SAB does not consistently exert a pro- or anti-regulation influence is an important factor in its institutional stability and credibility to the public.

There is still room for improvement in SAB's performance of routine reviews. Although many of SAB's activities are subject to public scrutiny, the process for selecting scientists to serve on SAB committees could benefit from more public discussion and reconsideration.[5] SAB has suffered from some degree of mistrust on this account.

The proper balance in use of standing versus ad hoc committees is also a difficult issue. The advantage of standing committees is that they provide some consistency and continuity in SAB reviews of complex issues. If SAB is to continue its historical practice of relying heavily on standing committees and subcommittees, it should consider making more aggressive use of ad hoc consultants than it has in the past. Such consultants—who were used to a limited extent in the reviews of formaldehyde, gasoline vapor, and nitrates—can better assure that SAB has the benefit of the best scientific talent available on the particular question before the agency.

SAB review of agency initiatives is already a fairly open process. The deliberations of the committees are typically in public view, and interested scientists are entitled to make written and oral comments before committees. The case studies of gasoline vapor and perchloroethylene, however, revealed that the committee's resolution of a particular issue is not always apparent at the public meeting. SAB's final letter to the administrator on gasoline vapor seemed to be at variance with the apparent consensus aired at the public meeting of the Environmental Health Committee. Similarly, the contents of the SAB letter to the administrator on perchloroethylene surprised agency scientists who had attended the Halogenated Organics Subcommittee's public review of the issue. Certainly, vote counting is not an appropriate procedure at SAB meetings, since some members will know more than others about a specific topic. It is nonetheless important to the credibility of the process to achieve closure in a public forum. The public closure principle may on occasion require an extra meeting, but the benefit of enhanced credibility would seem to justify the extra time.

One of the troubling characteristics of SAB's process is the relatively small amount of participation by scientists from environmental and public health groups. As the perchloroethylene case illustrates, public hearings tend to be dominated by scientists who represent companies or trade associations. In light of the asymmetry in public input, it is remarkable that SAB has sustained a reputation for objectivity.

This problem is not SAB's fault. There is no evidence that the lack of participation by environmentalists reflects distrust of the process. The more critical explanation is that such advocacy groups lack adequate access to the specialized scientific expertise that is necessary for effective participation. Most practicing scientists do not pursue careers in environmental advocacy groups, and such groups lack the resources necessary to hire academic consultants on the wide range of technical issues that SAB addresses. The few scientists employed by environmental groups are spread very thin. This problem will not be solved unless the scientific resources of the environmental community are expanded or the current allocations to litigators are diminished to allow more in-house scientific staff.

Over the years SAB has become sensitive to the criticism that it is used as a tool by potential regulatees to block or delay regulatory initiatives. For example, the General Accounting Office (GAO) produced a report in 1983 charging that lengthy SAB reviews of health-assessment documents were contributing to delays in the agency's air toxics program.[6] The delays documented in the chapter on perchloroethylene were of concern to GAO. In light of this criticism, SAB has become sophisticated about scrutinizing the rationales for requests to rereview a particular issue. Although rereviews can be justified on the basis of major new evidence or unresponsiveness by EPA, the case study of formaldehyde illustrates how the agency and SAB have become reluctant to grant marginal requests for rereview.

One of the difficult challenges for SAB is to focus on scientific matters and avoid becoming embroiled in questions of how a particular policy question should be resolved. On occasion SAB has been accused of addressing risk-management questions under the guise of scientific review. The case study of perchloroethylene revealed EPA and SAB skirmishing over the question of whether the chemical should be categorized as a probable or possible human carcinogen (B2 versus C). This is a case where assessment and management are difficult to separate, since the categorization decision tends to trigger particular regulatory responses by EPA offices and state agencies. In this type of instance, limiting SAB comments to purely scientific issues creates a dilemma for scientists who know but cannot comment on the regulatory consequences of their recommendations. In the perchloroethylene case EPA properly solicited comments from the board on science-policy issues. SAB is an advisory body, and expanding the dialogue between SAB and EPA can serve to clarify SAB advice without threatening SAB independence.

Taken as a whole, SAB must be considered a modestly successful contribution to resolving the regulatory science dilemma. SAB has provided EPA access to outside expertise by serving as a bridge between regulators and the scientific community. This bridge—and the stamp of scientific approval it offers—has earned EPA a badly needed measure of credibility. The EPA-SAB relationship has proven to be especially productive when committees develop a sound understanding of the workings of a specific program office. The SAB's experience under the Safe Drinking Water Act illustrates this point, as shown in the chapter on nitrates.

The major limitation on SAB's influence is the structure of the organization itself: a small in-house scientific staff, no capacity to generate original data, a limited ability to set its own agenda, reliance on poorly compensated scientists to review complex issues, and a high rate of turnover among EPA administrators and middle-level managers who have widely varying degrees of interest in science. SAB by itself is not powerful enough to influence whether or not frontier science is used in regulatory choices, but it can prevent EPA from making incompetent or poorly reasoned decisions. In the long run SAB may also play an increasingly influential role on the broader questions about EPA priorities for research and regulation.

THE CHEMICAL INDUSTRY INSTITUTE OF TOXICOLOGY

During the last ten years CIIT has established itself as a credible and reliable source of data on key toxicological questions. While CIIT scientists have long been respected in scientific circles, it was the formaldehyde controversy that caused the institute to come of age. When CIIT released the first definitive data in 1979 showing that formaldehyde causes nasal cancer in rats, it became apparent to everyone that this organization was prepared to place the interests of science and public health ahead of the short-run commercial interests of some of its sponsors. In short, CIIT has established a reputation for producing sound science.

One of CIIT's great frustrations has been the slowness of EPA to incorporate its mechanistic research into official risk-assessment reports. The case study of formaldehyde documented these frustrations. In fairness to EPA, it should be noted that CIIT's research on delivered doses of formaldehyde was at an evolutionary stage when it was originally reviewed by EPA risk assessors. Since the mid–1980s CIIT's work on formaldehyde has gained more widespread acceptance within the scientific community and regulatory agencies.[7]

The evolution of CIIT's formaldehyde research program illustrates a more fundamental evolution in CIIT's research mission. Originally, CIIT was conceived of as a vehicle for performing basic toxicity tests on chemicals in widespread use. More recently, CIIT has deemphasized standard toxicity testing and emphasized the development of mechanistic information that may permit valid extrapolations from high to low doses and from rodents to humans. The new emphasis on mechanistic research is motivated by the emergence of other testing programs (e.g., the National Toxicology Program) and the paucity of scientific knowledge about the standard mechanistic assumptions used in risk assessment. Since EPA's default assumptions about cancer mechanisms tend to be pessimistic, many scientists expect that acquisition of genuine mechanistic knowledge will tend to reduce EPA's estimated cancer risk estimates. CIIT is now using a risk-assessment framework to focus its research on those mechanistic assumptions that are crucial to risk assessment and regulatory decision making.

For the sponsors of CIIT, success entails more than the production of knowledge. CIIT needs to show that it can change the way the scientists and federal agencies look at risk assessments. In the gasoline vapor controversy CIIT demonstrated that it has begun to make this kind of difference. A small team of CIIT scientists showed in a convincing fashion that the tumors caused by experimental exposures to unleaded gasoline are caused by a mechanism that is unique to the male rat kidney. In the absence of this mechanistic information, EPA was prepared to regulate refueling emissions primarily on the basis of carcinogenicity. CIIT's data caused EPA to rethink the issue. In particular, the case study revealed how CIIT's data—in combination with sympathetic interpretations by Bernard Goldstein and the Health Effects Institute—had a profound effect on the thinking of EPA Administrator Lee Thomas. The change has not been as profound as CIIT's scientists would like. EPA's risk assessment still classifies unleaded gasoline as a "probable human carcinogen," although a range of risk estimates is reported. CIIT's mechanistic work on unleaded gasoline has nonetheless dispelled any lingering doubts about whether an industry-funded research organization can change the position of EPA on a key scientific issue that is relevant to regulation.

CIIT's research on male rat kidney tumors did have a demonstrable impact on EPA's water quality standard (MCL) for paradichlorobenzene (PDCB). Before CIIT's research was released, PDCB was classified by EPA as a probable human carcinogen (B2). After CIIT's research was considered, PDCB was reclassified as a possible human carcinogen (C). Although EPA originally considered a MCL of 0 for PDCB, the final MCL was set at .075 ppb in light of the new science.[8]

More thought needs to be given to how CIIT's research findings should be marketed to federal agencies. The traditional view has been that CIIT should publish its results in peer-reviewed scientific journals and let the data speak for themselves. The flaw in this strategy is that such data may be ignored or misinterpreted by federal agencies. CIIT has taken a step further and developed in-house expertise in risk-assessment modeling in order to show how its data might be used in risk assessment.[9] In the case of formaldehyde, CIIT scientists may have been too aggressive in promoting their work, since they ultimately developed an adversarial relationship with certain key risk assessors in the federal government. CIIT scientists cannot afford to develop the "government-bashing" reputation that limits the effectiveness of scientists from trade associations. Even if CIIT educates the critical trade associations on its findings and encourages them to carry the burden of scientific advocacy, it is not clear that these associations have scientific credibility within federal agencies. The American Industrial Health Council has maintained some credibility on these issues by avoiding chemical-specific controversies. The question of how CIIT scientists should participate in the government's risk-assessment proceedings raises important managerial issues. There is a serious trade-off between laboratory productivity and extensive participation in the risk-assessment process. Moreover, the professional skills required to market mechanistic research are probably different from those required to produce such information in the laboratory, which may have been part of CIIT's problem in the case of formaldehyde.

The case study of gasoline vapor suggests that CIIT will be most influential when it can persuade key scientists from the academic community that EPA should utilize CIIT's research. The implication is that CIIT can benefit from strong ties to university-based programs in toxicology and risk assessment.

The lack of an institutional relationship between CIIT and EPA may place some limits on how influential CIIT's research can be, at least in the short run. SAB can play some mediating role, but its influence on EPA is limited. Only recently has EPA begun to build a team of scientists that can understand and utilize mechanistic information. It may be wise for CIIT to include some first-rate EPA scientists on its advisory panels.

The timing of CIIT's research is also a critical issue. In order to have maximum societal impact, CIIT needs to produce data at the early stages of a regulatory investigation. Once a corporation or federal agency takes a tentative stance on key scientific issues, the burden of persuasion for new science is heightened significantly. Hence CIIT needs to set its research priorities with an eye toward the timing of corporate and regulatory decisions—uncertain as those schedules may be. The implication is that CIIT must be cautious about making research investments on scientific issues that are too immature (i.e., knowledge is lacking about what assumptions will be critical) or too mature (i.e., the agency or industry is already too committed to turn back on a key assumption). As long as CIIT can produce some clear successes on salient issues (e.g., gasoline vapor), it will retain the freedom to address some of the more critical long-run methodological questions. Research aimed at improving risk assessment generally may have broader acceptance than defensive projects aimed at undercutting EPA's regulatory ambitions.

Although CIIT has had a remarkable impact for a small organization, it cannot currently address a broad range of scientific issues due to limited resources. The number of corporate sponsors of CIIT declined in the early 1980s but is now on the increase again. In order to expand its industrial base, CIIT needs to dispel the myth that it addresses only the concerns of chemical companies. The organization's emphasis on mechanistic issues in risk assessment is relevant to virtually every major manufacturing industry in America, including pharmaceuticals, energy/petroleum, metals, and foods.

A more controversial question is whether CIIT should pursue public-sector support for its research. Given CIIT's relatively successful track record to date, this kind of major reform should be pursued with caution. The dangers of developing dependence on the government for project support include a weakened ability to marshal multidisciplinary teams on urgent questions and a fragmentation in the organization's research priorities. If CIIT could persuade the federal government to provide an unrestricted block grant for mechanistic research on risk assessment, the case for partnership would be much stronger.

One of the key managerial challenges at CIIT is to fashion disclosure policy in a way that provides some room for sponsor oversight but does not raise competitiveness issues. CIIT must, of course, continue its policy of providing early disclosure of results to government agencies and the public at large. That policy served CIIT handsomely in the case of formaldehyde. The case study of gasoline vapor discussed a creative arrangement that permitted certain scientists from the petroleum industry to monitor work in progress without undermining CIIT's nondisclosure policy. If this kind of arrangement can be continued in the future without creating competitiveness issues or compromising CIIT's independence, CIIT may find it easier to persuade companies and trade associations to provide larger contributions for project support. As CIIT has moved into more generic, mechanistic issues, the sensitivity of disclosure issues has diminished.

On the whole, CIIT must be considered a successful influence in America's effort to resolve the regulatory science dilemma. The beauty of the CIIT model is that it allows companies to pool resources into an organization that produces more scientific impact than those same dollars would if they were spread across the contributing companies. Perhaps more surprisingly, CIIT has shown that it is possible to create a credible industry-funded scientific organization to address toxic chemical issues. CIIT can provide expert knowledge to EPA on key regulatory issues without raising questions about whether EPA has delegated its power to unaccountable industrial experts. EPA scientists, aided by the Science Advisory Board, can review CIIT research along with other evidence that comes to the agency's attention. The key challenge for CIIT is to demonstrate to its current and future sponsors that it can produce science that will consistently improve the scientific basis of corporate and regulatory decisions.

THE HEALTH EFFECTS INSTITUTE

When one considers the history of EPA regulation of the auto industry, the Health Effects Institute must be considered a revolutionary concept. Throughout the 1970s EPA and Detroit fought tooth and nail over what the health consequences of auto emissions were, what kinds of emission reductions were technically feasible, how much emission standards would cost, and what regulatory policies were in the public interest. The disagreements were not just about values; each side had its own version of the facts. Detroit did not trust EPA's scientific work, EPA was skeptical about industry-supported science, and many first-rate scientists avoided the field. Meanwhile, the public interest was poorly served because little progress was made toward a genuine understanding of how emissions from motor vehicles affect public health.

Charles Powers and Michael Walsh saw an opportunity in this destructive war between EPA and Detroit. The opportunity was to hasten peace—at least on the scientific front—by creating a new scientific organization characterized by independence, integrity, and impeccable expertise. The Health Effects Institute, a partnership of the public and private sectors, would become both the arbiter of scientific disputes and a catalyst for the development of a mature body of scientific knowledge.

In the face of such lofty objectives, what can we say now after ten years of experience with HEI? Obviously, it is too early to make a definitive evaluation, but the preliminary signs are encouraging. The scientists who have participated in the HEI experiment talk about it with a remarkable sense of excitement. Among those outsiders who know HEI, the predominant view is that HEI is an objective, high-quality research organization. Some would argue that HEI got off to a slow start and has not yet been extremely productive. Remember that HEI does not have an in-house data generation capability and thus relies on others to do the original research. The HEI Research Committee has shown that it can identify critical research needs and persuade first-rate scientists throughout the world to work on them. HEI has also demonstrated an innovative process of peer review that results in a published statement by the Review Committee on the contributions and limitations of each sponsored study. The ultimate test of HEI's research—whether it can clarify the impact of air pollution on human health—must be assessed within a longer time frame.

HEI has shown that it can serve as a credible interpreter of sensitive data. The case study of gasoline vapor traced how HEI became involved in its first regulatory controversy. The key regulatory offices within EPA were determined to regulate refueling emissions and were making quantitative estimates of cancer risk to buttress their case. Although HEI had not sponsored any of the original research on the carcinogenic risks of inhaling gasoline vapors, the Board of Directors accepted Detroit's request—and a belated EPA concurrence—to evaluate the scientific evidence linking unleaded gasoline vapors to human cancer risk.

As the case study reveals, the Health Effects Institute's review had a powerful influence on the thinking of EPA Administrator Lee Thomas. The fact that EPA was a cosponsor of HEI was an important factor in Thomas's decision to delay a final decision on regulating refueling emissions. After a meeting with HEI scientists, Thomas became convinced that the agency's cancer risk estimates were far more fragile than he had been led to believe. Interestingly enough, CIIT's mechanistic research on the toxicity of unleaded gasoline was a major influence on the thinking of HEI scientists. It should be noted, however, that it was HEI, not CIIT, that ultimately obtained a personal briefing of the EPA administrator on the issue during the critical deliberative period.

HEI's intervention caused Thomas to rethink the entire refueling issue—a process that ultimately consumed another eighteen months. Thomas felt that HEI raised scientific issues that SAB and his staff had not emphasized, although others argued that Thomas had not fully grasped his staff briefings on this question. In any event, HEI made a difference. Cancer risk was ultimately deemphasized as a rulemaking rationale, and ozone control was made the central consideration.

A case can be made that HEI entered the gasoline vapor controversy too late to have maximum impact. The late entrance also generated conflict with EPA officials that could have been avoided by an earlier intervention. HEI appears to have become more sensitive to the timing issues. For example, a more recent HEI review of the potential health effects of methanol was published well before an EPA administrator was facing a controversial regulatory call.[10] Now that the Bush administration appears to be advocating increased use of alcohol- and methanol-based fuels, HEI has positioned itself to have a more profound influence on the ensuing regulatory science discussions.

The early signs also indicate that some of the original data generated by HEI are both of high quality and of direct relevance to EPA decision making. For example, HEI has recently completed a major multicenter study of the effects of carbon monoxide on angina among exercising heart patients.[11] As we saw in Chapter 9, this study was undertaken with remarkable rigor and is likely to be the centerpiece of EPA's upcoming review of the national ambient air quality standard for carbon monoxide. Although some of HEI's research has been criticized for lack of relevance to sponsor needs, the multicenter study of CO and the reviews of gasoline vapor and methanol indicate that HEI is often responsive to the requests of its sponsors.[12]

In light of HEI's scientific and political clout, it is disappointing that HEI is not a more secure, thriving organization. Until very recently, HEI's budget had remained roughly flat (in nominal terms) since it was created. Efforts to expand HEI's mission to encompass other regulatory science issues have met with only modest success (e.g., a new project on asbestos), although an exciting effort to add an epidemiological program at HEI is now under consideration. The HEI model has not yet proliferated into other realms of regulatory science, even though the social strategy that it reflects could be applied fruitfully to other industries such as chemicals, petroleum, and pharmaceuticals.

HEI has been only a qualified success in part because some key career officials in EPA resent HEI. Some scientists in ORD, for example, believe that HEI's budget was created at the expense of their own budget. Regulators in OMS and OAQPS were not pleased when HEI threw a monkey wrench into their efforts to persuade Lee Thomas to sign off on a major new regulation of refueling emissions. The founders of HEI offended many EPA scientists by intimating— and in some cases declaring—that first-rate health effects research could not be performed within EPA. From the very beginning, HEI was launched through high-level negotiations, and many public servants within EPA were not privy to the discussions.

The relationship between HEI and parts of EPA was so weak in the mid–1980s that a clumsy attempt was made at one point to eliminate EPA funding of HEI. The HEI Board of Directors blocked this effort by persuading the appropriations committees in Congress to insist on continued funding. The unexpected feature of the HEI experience in the 1980s has been the degree of tension between EPA and HEI (rather than the expected tension between EPA and Detroit). In recent years, however, it appears that the HEI-EPA relationship has begun to improve.

Another plausible explanation for HEI's limited success is that the organization is unfamiliar, inaccessible, and somewhat elitist. According to Ellen Silbergeld of the Environmental Defense Fund, "HEI has been a clubby group of academics from Cambridge who don't participate in the nuts and bolts of environmental science and public policy."[13] While this may be an overstatement, it may also reflect a germ of truth. Bernard Goldstein, an advocate of HEI, has noted that the HEI Board of Directors would be a more credible advocate of its own cause if it were also in the trenches making a more general case in favor of expanded funding for environmental science research.[14]

The original HEI Board of Directors (Archibald Cox, Donald Kennedy, and William Baker) gave HEI the integrity and credibility that was necessary to get started and prevent early death. The scientists on the HEI Research and Review Committees are among the most distinguished researchers in the nation. They provide HEI with an immediate source of scientific expertise and credibility. Each year another dozen or so scientists are added to the list of researchers who have received research grants from HEI. But aside from these individuals and a very few consultants, very few people have experienced the excitement of HEI, and few opinion leaders in America know about it.

If the HEI model is to expand or proliferate, its advocates will need to modify their ivory-tower image and rub shoulders with the rest of the environmental science community. HEI's large annual meeting of scientists has been a useful step in this direction. An expanded, revitalized Board of Directors would be an even more important step.

HEI also suffered from some instability in managerial leadership in the 1980s, although there has been significant continuity among the in-house and consulting scientists. There have been four executive directors in ten years, a track record that does not permit cultivation of sound professional relationships with key actors in Congress, EPA, the states, various industries, the mass media, and other subcultures that are fertile territory for promoting the HEI model. Continuity in the executive director position would be less critical if the HEI Board of Directors participated more actively in the environmental science and policy-making communities.

Overall, the Health Effects Institute is perhaps the most innovative approach to addressing the regulatory science dilemma that has yet been attempted. The organization, while less secure than CIIT and SAB, has survived the pains of birth and begun to make some significant scientific contributions. Despite these encouraging signs, one cannot escape the conclusion that HEI, and the model it represents, is an underutilized national resource.

THE NEW ENVIRONMENTAL POLITICS

In most arenas of environmental policy making, the burden of scientific analysis lies with regulators to make a case that regulation is required to protect public health. Although regulators are not required to eliminate uncertainty before they regulate, the burden is on them to demonstrate through risk assessment that human health may be endangered by toxic chemical exposure.

During the last twenty years environmentalists have become frustrated with this paradigm of environmental policy making. They argue that when the government shoulders the burden of proof, society gets too much analysis and not enough reduction in human exposure to toxic chemicals. The burden of scientific analysis, it is argued, should be shifted from the government to the organizations that expose people to toxic substances.

Those who advocate such a shift in the burden of proof are gaining the upper hand in the battleground of environmental politics. California's Proposition 65, which requires zero discharges into water and warning provisions for other exposures to toxic chemicals, embodies the new allocation of responsibility. Organizations may continue to emit toxic substances only if they can prove that their emissions do not pose a significant risk of cancer or reproductive effects. Proposals to facilitate mass toxic tort suits are also gaining more attention in policy debates.

The case study of nitrates provided insight into the impatience of key congressional staffers. Under the Safe Drinking Water Act the trend is toward ambitious legislative requirements for standard setting with fixed deadlines. Frequently, Congress does not provide the resources necessary to meet these deadlines. EPA and SAB are only beginning to grapple with the logistical implications of this agency-forcing strategy. The debate in Washington about air toxics legislation has also begun to move in this direction. A consensus seems to have emerged that best available control technology should be applied to a large number of industrial sources even if EPA cannot establish that such sources constitute a significant threat to public health. While less explicit in this case, the burden of proof seems to be shifting to industry.

The new dynamic of environmental politics, if it is permanent, has profound implications for the regulatory science dilemma. What it means is that uncertainties in risk assessment will work not to complicate the case for regulation but rather to prevent an emitter from proving that his emissions are "safe." In short, private-sector organizations that are responsible for generating human exposures to toxic chemicals are going to need a more sophisticated capability to monitor and evaluate such exposures.

Since many firms lack the resources and expertise to produce risk assessments, the demand for regulatory science organizations such as CIIT and HEI is likely to proliferate. In California, for example, the Institute for Evaluating Health Risk (IEHR) is being launched by some of the same entrepreneurs who created HEI. Although the mission of IEHR is still somewhat ambiguous, it is intended to provide both the public and private sectors with the capacity to produce credible and timely risk assessments under the terms of Proposition 65. If the Proposition 65 model spills over into other states, it is likely that IEHR-type organizations will multiply.

In the final analysis, we should not expect the regulatory science dilemma to be resolved until scientists begin to understand more about the basic mechanisms of chemical toxicity. Without an understanding of mechanisms, risk assessors must continue to make unverifiable extrapolations from high to low doses and from animals to humans. Our society needs an expanded, long-term commitment to producing knowledge about the effects of human exposures to toxic chemicals. An expanded resource base for science, by itself, is not enough. Creative institutional relationships must be fostered that will facilitate both scientific credibility and responsiveness to societal needs. Organizations such as SAB, CIIT, and HEI, despite their limitations, have just this kind of character and should be nurtured and fostered in the years ahead.

NOTES

1. T. F. Yosie, "EPA's Risk Assessment Culture," *Environmental Science and Technology*, vol. 21, 1987, p. 529.

2. Discussion at the Belmont Workshop on Scientific Advice for Environmental Health Regulation, Elkridge, Maryland, October 1988.

3. U.S. Environmental Protection Agency, *Unfinished Business: A Comparative Assessment of Environmental Problems*, U.S. EPA Office of Policy Analysis, Office of Policy Planning and Evaluation, Washington, D.C., 1987.

4. U.S EPA, *Future Risk: Research Strategies for the 1990's*, SAB–EC–040, Washington, D.C., September 1988.

5. SAB seeks nominations from scientific societies, the National Academy of Sciences, and environmental and industry groups. The kinds of qualifications sought are published in the *Federal Register*. Ultimately, the SAB director works with the SAB Executive Committee members and the relevant committee chair before making a recommended appointment to the EPA administrator. See T. S. Burack, "Of Reliable Science: Scientific Peer Review, Federal Regulatory Agencies, and the Courts," *Virginia Journal of Natural Resources Law*, vol. 7, 1987, p. 43. Also see *The Mission and Functioning of the EPA Science Advisory Board*, Report to the Board, October 23, 1989.

6. U.S. General Accounting Office, *Delays in EPA's Regulation of Hazardous Air Pollutants*, U.S. Comptroller General's Office, Washington, D.C., 1983.

7. Discussion at the Belmont Workshop on Scientific Advice for Environmental Health Regulation, Elkridge, Maryland, October 1988.

8. U.S. Environmental Protection Agency, *Federal Register*, vol. 52, July 8, 1987, pp. 25694–96.

9. See, for example, T. B. Starr and R. D. Buck, "The Importance of Delivered Dose in Estimating Low-Dose Cancer Risk from Inhalation Exposure to Formaldehyde," *Fundamental and Applied Toxicology*, vol. 4, 1984, pp. 740–53.

10. Health Effects Institute, *Automotive Methanol Vapors and Human Health: An Evaluation of Existing Scientific Information and Issues for Future Research*, May 1987.

11. Health Effects Institute, *Acute Effects of Carbon Monoxide Exposure on Individuals with Coronary Artery Disease*, Research Report Number 25, 1989.

12. Thomas Grumbly, personal communication, 1988.

13. Ellen Silbergeld, personal communication, January 1989.

14. Bernard Goldstein, personal communication, January 1989.

APPENDIX

BELMONT WORKSHOP PARTICIPANTS

Dr. Donald G. Barnes
Acting Director, Science Advisory Board
U.S. Environmental Protection Agency (A–101)
401 M Street, S.W.
Washington, DC 20460
Tel.: (202) 382–4126

Mr. Richard M. Cooper
Williams and Connolly
839 17th Street, N.W.
Washington, DC 20006
Tel.: (202) 331–3017

Ms. Elizabeth Drye
U.S. Environmental Protection Agency
Office of Policy Analysis, PM–220
Room 3003
401 M Street, S.W.
Washington, DC 20460
Tel.: (202) 382–2730

Ms. Susan Egan-Keane
Abt Associates
55 Wheeler Street
Cambridge, MA 02138
Tel.: (617) 492–7100, X5397

Dr. William Farland
Acting Director
Office of Health and Environmental Assessment
U.S. Environmental Protection Agency
401 M Street, S.W. (RD 689)
Washington, DC 20460
Tel.: (202) 382–7315

Mr. Paul Gilman
Administrative Assistant
Senator Dirksen Building
Room 434
Washington, DC 20510
Tel.: (202) 244–6621

Dr. John D. Graham
Associate Professor of Policy and Decision Sciences
Harvard School of Public Health
Department of Health Policy and Management
677 Huntington Avenue
Boston, MA 02115
Tel.: (617) 732–1090

Mr. Thomas Grumbly
President
Clean Sites, Inc.
1199 N. Fairfax Street
Suite 400
Alexandria, VA 22314
Tel.: (703) 739–1240

Dr. Fred Hoerger
Regulatory and Policy Consultant
Dow Chemical Company
1803 Building
Midland, MI 48674
Tel.: (517) 636–0682

Dr. Donald F. Horning
Chairman
Department of Environmental Science and Physiology
Harvard School of Public Health (1–1411)
665 Huntington Avenue
Boston, MA 02115
Tel.: (617) 732–1272

Dr. Richard L. Klimisch
Executive Director

Environmental Activities Staff
General Motors Corporation
General Motors Technical Center
30400 Mound Road
Warren, MI 48090–9015
Tel.: (313) 974–0065

Dr. William Lowrance
Senior Fellow and Director
Life Sciences and Public Policy Program
The Rockefeller University
1230 York Avenue
New York, NY 10021–6399
Tel.: (212) 570–8679

Dr. Judith A. MacGregor
Manager
Chevron Environmental Health Center
P.O. Box 4954
Richmond, CA 94804–0054
Tel.: (415) 231–6040

Dr. Roger McClellan
President
Chemical Industry Institute of Toxicology
6 Davis Drive
Research Triangle Park, NC 27709
Tel.: (919) 549–8201

Dr. Robert Neal
Adjunct Professor of Biochemistry
Center in Molecular Toxicology
Department of Biochemistry
Vanderbilt University School of Medicine
Medical Center North
Nashville, TN 37232–2745
Tel.: (615) 322–2261

Dr. Norton Nelson
Professor of Environmental Medicine
NYU Medical Center
550 First Avenue
New York, NY 10016
Tel.: (914) 351–2566

Dr. Paul R. Portney
Director

Center for Risk Management
Resources for the Future
1616 P Street, N.W.
Washington, DC 20036
Tel.: (202) 328–5093

Mr. Charles Powers
Partner
Resources for Responsible Management
264 Beacon Street
Boston, MA 02116
Tel.: (617) 266–7622

Ms. Susan Putnam
Candidate, Sc.D.
Harvard School of Public Health
Department of Health Policy and Management
677 Huntington Avenue
Boston, MA 02115
Tel.: (617) 732–1090

Mr. Alon Rosenthal
Candidate, Sc.D.
Harvard School of Public Health
Department of Health Policy and Management
677 Huntington Avenue
Boston, MA 02115
Tel.: (617) 732–1090

Mr. Eric Ruder
Environ Corp.
210 Carnegie Center, Suite 201
Princeton, NJ 08540
Tel.: (609) 452–9000

Dr. Robert Scala
Senior Scientific Advisor
Exxon Biomedical Sciences Inc.
Mattlers Road, CN 2350
East Millstone, NJ 08875–2350
Tel.: (201) 873–6061

Dr. James Senger
Vice President, Environmental Policy Staff
Monsanto Company
800 N. Lindbergh Boulevard/A3NA
St. Louis, MO 63167
Tel.: (314) 649–8873

Dr. Ellen Silbergeld
Chief Scientist
Environmental Defense Fund
1616 P Street, N.W.
Washington, DC 20036
Tel.: (202) 387–3500

Mr. Larry Slimak
Executive Secretary
Motor Vehicle Manufacturers Association
300 New Center Building
Detroit, MI 48202
Tel.: (313) 872–4311

Dr. James Swenberg
Head, Department of Biochemical Toxicology and Pathobiology
Chemical Industry Institute of Toxicology
P.O. Box 12137
6 Davis Drive
Research Triangle Park, NC 27709
Tel.: (919) 541–2070

Dr. Arthur R. Upton
Professor and Director
Institute of Environmental Medicine
NYU Medical Center
550 First Avenue
New York, NY 10016
Tel.: (212) 340–5280

Dr. Terry R. Yosie
Vice President
American Petroleum Institute
1220 L Street, N.W.
Washington, DC 20005
Tel.: (202) 682–8090

SELECT BIBLIOGRAPHY

Brain, Joseph D., Barbara D. Beck, A. Jane Warren, and Rashid A. Shaikh, eds.
    Variations in Susceptibility to Inhaled Pollutants. Johns Hopkins University Press,
    Baltimore, Md., 1988.

Crandall, Robert W. *Controlling Industrial Pollution*. Brookings Institution, Washington,
    D.C., 1983.

Graham, John D., Laura Green, and Marc J. Roberts. *In Search of Safety: Chemicals
    and Cancer Risk*. Harvard University Press, Cambridge, Mass., 1988.

Hart, Ronald W., and Fred D. Hoerger, eds. *Carcinogen Risk Assessment: New Directions
    in the Qualitative and Quantitative Aspects*. Banbury Report 31. Cold Spring
    Harbor Laboratory, N.Y., 1988.

Jasanoff, Sheila. *The Fifth Branch of Government*. Harvard University Press, Cambridge,
    Mass., 1990.

Lave, Lester B., ed. *Quantitative Risk Assessment*. Brookings Institution, Washington,
    D.C., 1982.

Lowrance, William W. *Modern Science and Human Values*. Oxford University Press,
    New York, 1985.

Mendeloff, John M. *The Dilemma of Toxic Substance Regulation*. MIT Press, Cambridge,
    Mass.,1988.

Portney, Paul R., ed. *Public Policies for Environmental Protection*. Resources for the
    Future, Washington, D.C., 1990.

Whipple, Chris, ed., *De Minimis Risk*. Plenum Press, New York, 1987.

INDEX

ABOUT THE CONTRIBUTORS

ELIZABETH DRYE is a staff member in the Office of Policy Analysis, U.S. Environmental Protection Agency.

SUSAN EGAN-KEANE is an environmental analyst with Abt Associates, Inc., in Cambridge, Massachusetts.

JOHN D. GRAHAM is Associate Professor of Policy and Decision Sciences in the Department of Health Policy and Management at the Harvard School of Public Health.

THOMAS P. GRUMBLY is President of Clean Sites, Inc., and former Executive Director of the Health Effects Institute.

DAVID HOLTGRAVE is Assistant Professor of Family Medicine at the Health Sciences Center, University of Oklahoma, Oklahoma City, Oklahoma.

ROBERT A. NEAL is Adjunct Professor of Biochemistry in the Center in Molecular Toxicology, Vanderbilt University School of Medicine, and former President of the Chemical Industry Institute of Toxicology.

SUSAN W. PUTNAM is a doctoral candidate in health policy and management at the Harvard School of Public Health.

ALON ROSENTHAL is Lecturer at the Harvard School of Public Health.

ERIC RUDER is a staff scientist at Environ Corporation in Princeton, New Jersey.

TERRY F. YOSIE is Vice President, Health and Environment, of the American Petroleum Institute and former Director of the U.S. Environmental Protection Agency's Science Advisory Board.